Priorities in Biopesticide Research and Development in Developing Countries

CABI *Bioscience* is a division of **CAB** *International*, an inter-governmental, not-for-profit, mission-oriented organization dedicated to improving human welfare world-wide through the dissemination, application and generation of scientific knowledge in support of sustainable development. Emphasis is placed on agriculture, forestry, human health and the management of natural resources and particular attention is given to the needs of developing countries.

CABI *Bioscience*'s **Biopesticides Programme** is committed to the development and use of biopesticides as safe, environmentally friendly alternatives to chemical pesticides. The Programme carries out collaborative inter-disciplinary research and development, offers training in insect pathology, runs the International Biopesticide Consortium for Development (IBCD), disseminates information and promotes the role and value of biopesticides in sustainable crop production, poverty alleviation and wealth generation.

Biopesticides series

1. *Chemical Pesticide Markets, Health Risks and Residues*
   J. Harris

2. *Priorities in Biopesticide Research and Development in Developing Countries*
   J. Harris and D.R. Dent

Biopesticides are biological pesticides based on beneficial insect and weed pathogens and entomopathogenic nematodes. Pathogens used as biopesticides include fungi, bacteria, viruses and protozoa. Produced, formulated and applied in appropriate ways, such biopesticides can provide ecological and effective solutions to pest problems.

The aims of the Biopesticides series are to more widely appraise and promote the role and value of biopesticides as alternatives to chemical pesticides and to improve awareness of the opportunities offered by biopesticides.

The series has been developed by the Biopesticides Programme at CABI *Bioscience* as part of its mission to disseminate information and promote the role and value of biopesticides.

Priorities in Biopesticide Research and Development in Developing Countries

Biopesticides Series No. 2

Jeremy Harris and David R. Dent

CABI Bioscience (UK Centre)
Ascot, UK

CABI *Publishing*

CABI Publishing is a division of CAB International

CABI Publishing
CAB International
Wallingford
Oxon OX10 8DE
UK

Tel: +44 (0)1491 832111
Fax: +44 (0)1491 833508
Email: cabi@cabi.org
Web site: www.cabi-publishing.org

CABI Publishing
875 Massachusetts Avenue
7th Floor
Cambridge, MA 02139
USA

Tel: +1 617 395 4056
Fax: +1 617 354 6875
Email: cabi-nao@cabi.org

A catalogue record for this book is available from the British Library, London, UK
A catalogue record for this book is available from the Library of Congress, Washington DC, USA

ISBN 0 85199 479 2

First published 2000
Transferred to print on demand 2005

Printed and bound in the UK by Antony Rowe Limited, Eastbourne.
=

Contents

Preface

Biological pesticides based on pathogenic micro-organisms specific to their target pest offer an ecological and effective solution to pest problems. Such biopesticides are an alternative to chemical pesticides that continue to be used inappropriately, particularly in developing countries, threatening the environment and human health. However, despite the enormous potential of biopesticides as substitutes for chemical pesticides and for use in IPM programmes, their development, commercialisation and use has not yet lived up to expectations.

In the first survey to attempt to identify the principle reasons for this relative lack of development, over 1000 biopesticide researchers in developing countries were sent a questionnaire requesting information about their research activities, funding, levels of technical expertise and availability of equipment. Their opinions were also sought on priorities for funding biopesticide research and development, the constraints to successful biopesticide development and the barriers that impede commercial exploitation of potential biopesticide products.

The response rate of 12% from 30 countries was excellent for an unsolicited mail-shot questionnaire of this type and is further evidence for the desire in developing countries to develop effective biopesticide products. The results of the survey confirmed that the principle difficulties and constraints that are facing researchers in developing countries involved in biopesticides relate to a lack of expertise in the crucial later stages of development. The results give further support to previous observations that biopesticide research is receiving mostly low investment, largely from the public sector and rarely involves the multi-disciplinary expertise required to develop a biopesticide from start to finish. The report concludes that targeted assistance in the form of specialised support, facilities and access to expertise on a multi-national, multi-institute, multi-disciplinary basis is required in developing countries in order to remove the constraints to the future successful development and utilisation of biopesticides.

Introduction

Despite decades of warnings, the inappropriate use of chemical pesticides continues to pose threats to the environment and human health, especially in developing countries. There have been massive upsurges in pesticide use in recent years (Harris, 2000) and increasing use, and often misuse, has led to increased problems of insecticide resistance. In addition, there are the problems of destruction of beneficial insects and other non-target organisms, toxic residues on produce, large stocks of obsolete pesticides requiring disposal and human poisoning. The World Health Organisation estimated that there are 25 million cases of acute occupational pesticide poisoning in developing countries and 20,000 deaths world-wide each year (Jeyaratnam, 1990). New chemicals with improved properties are available but are beyond the means of many farmers in developing countries.

The use of biological pesticides, based on micro-organisms which are highly specific to the target pest, has been proposed for many years. Biopesticides are biological pesticides based on beneficial insect and weed pathogens and entomopathogenic nematodes. Pathogens used as biopesticides include fungi, bacteria, viruses and protozoa. If produced, formulated and applied in appropriate ways, such biopesticides can provide ecological and effective solutions to pest problems. Biopesticides have been shown to be effective in controlling pests that have developed resistance to chemical pesticides, leave little or no toxic residues and are generally harmless to beneficial insects and other non-target organisms. A major benefit is that they are safe for use by humans and they represent a very much reduced hazard in terms of disposal and reuse of packaging than chemical pesticides. However, despite the enormous potential of biopesticides as substitutes for chemical pesticides and for use in IPM programmes, their development, commercialisation and use has not yet lived up to expectations.

The market for biopesticides is growing but still only represents less than 1% of the total crop protection market and most of this is accounted for by *Bt* based products (Lisansky, 1997). At the present time there are estimated to be 185 biopesticide products world-wide (Copping, 1998) (72 bacteria, 47 fungi, 40 nematodes, 24 viruses and 2 protozoa). Various reasons are given for the lack of product development and market penetration of biopesticides including: expectations that the multi-national agrochemical companies would take a lead in product development; over investment from "venture capital" initiatives (Waage, 1999); and pursuance of an inappropriate model for biopesticide development based on small research teams lacking the multi-disciplinary expertise required (Dent, 1997). The continued development of biopesticides in the future will be largely dependent on the public sector as the multi-national private sector seeks better investment prospects from biotechnology (Waage, 1999).

The expertise required to develop a biopesticide includes:

- Exploration, identification and screening of pathogen isolates
- Mass production
- Storage
- Formulation
- Application
- Ecology
- Toxicology and ecotoxicology
- Registration
- Commercialisation

Successfully identifying, and removing, the constraints to the commercialisation of biopesticides has been one of the features of recent work by the LUBILOSA programme and has resulted in the transfer of a product for the control of locusts to the private sector (see below). Many more opportunities exist where research results are ready to be implemented but relatively minor obstacles to implementation need to be correctly identified and tackled. As part of an initiative by CABI *Bioscience*, a questionnaire was constructed and sent to biopesticide researchers in developing countries to help identify the constraints to successful biopesticide development and the barriers that impede commercial exploitation of potential biopesticide products. The information gained from the response to this questionnaire is presented in this report and it is hoped that this can be used to help with the future prioritisation of areas for funding and assistance in developing biopesticides.

LUBILOSA, a case study

There is a recent precedent for the successful development of a mycopesticide in Africa which addresses many of the product development issues which apply to biopesticides generally. This is the LUBILOSA (**Lu**tte **Bi**ologique contre les **Lo**custs et **Sa**uteriaux) Programme, which has developed a product based on the African fungus, *Metarhizium anisopliae*, for control of desert locusts and other species of locusts and grasshoppers.

LUBILOSA is a collaborative programme between CABI *Bioscience*, IITA, GTZ, CILSS and a number of national programmes. Managed by CABI *Bioscience*, LUBILOSA has been supported by a range of development assistance agencies, including CIDA, DFID, NEDA, SDC, and USAID. LUBILOSA also works with the pest control industry to ensure effective registration and commercialisation of the product.

Over a period of 10 years, the team have taken an R&D concept — the formulation of fungal spores in oil — and turned it into a commercial product for the control of a wide range of locusts and grasshoppers. The mycoinsecticide is effective, economic and has minimal impact upon the environment. From this development has arisen new technologies for the mass production and storage of fungal pathogens, their formulation as ultra-low-volume sprays for ground and aerial application, and recommendations for their effective use as pest control agents which, unlike agrochemicals, persist in the environment to form a part of the ecology of the pest.

The LUBILOSA team has expertise in strain isolation, identification, characterisation and selection, its production, storage, formulation, application, registration and commercialisation. All of these steps are essential to ensure that a biopesticide is effective and commercially viable, meeting the needs and standards of all stakeholders. The team has demonstrated that the oil-based formulation of spores, applied at rates of 0.5 litres per hectare, can be used over large areas to control locusts and grasshoppers in the Sahel, East Africa and South Africa.

LUBILOSA has developed high quality production procedures and standards for the fungus. It has carried out economic studies of low input production and marketing systems to assess the optimal means of commercialisation, and has worked in partnership with industrial companies to ensure the effective licensing and production of the product. The product, trademarked "Green Muscle", was registered for use in South Africa in 1998 and registration is being sought in more African countries in the sub-Saharan region. The product is manufactured by two companies, Biological Control Products (South Africa) and Natural Plant Protection (France).

Many developing countries have research expertise in biopesticides and an adequate level of technology. However, as in many developed countries, there is a gulf between great research ideas and the implementation and adoption of a new technology. Biopesticide research is usually funded piecemeal, largely by the public sector and rarely involves multi-disciplinary teams that develop a microbial insecticide from start to finish (Dent, 1997). The general knowledge base in biopesticides is built up in a haphazard way, through the uncoordinated efforts of many scientists all pursuing their own individual research objectives and interests. This contrasts markedly with the more focused factory-like screening and development process which characterises agrochemical R&D that produces new chemical pesticides (Dent, 1997).

Questionnaire Design and Target Researchers

An effort was made to structure and word the questionnaire in a clear and easily understandable format for the target researchers in developing countries whose first language was highly unlikely to be English. The questionnaire was designed in consultation with Dr H. De Groote, socio-economist at IITA, Cotonou, Benin and drafts circulated amongst staff and students at CABI *Bioscience* to ensure that the questionnaire met these criteria. The final version that was sent out to biopesticide researchers forms Appendix 3. In recognition of their assistance, all respondents were promised a copy of this report.

The questionnaire was sent to 1038 researchers in biopesticides located in 50 developing countries whose contact details were obtained from the following sources:

1. 924 researchers listed in *Biopesticides: World-wide Directory of Research and Researchers* (R. Quinlan, 1995, CPL Press) — compiled from a survey of scientific papers published between 1992 and 1994
2. 99 researchers listed in the delegate list from the 1998 Society of Invertebrate Pathologists conference, Sapporo, Japan
3. 11 researchers listed in the delegate list from 1998 British Mycological Society conference, Southampton, UK
4. 4 researchers listed in the delegate list from 1997 BCPC symposium, Warwick, UK

Each of these sources was checked through to avoid duplicate entries when compiling the list of contacts. Researchers collaborating on CABI *Bioscience* projects such as LUBILOSA were not included on the mailing list. The questionnaires were sent via the Royal Mail who would where possible return those which did not reach their destination. As the questionnaires were returned, all completed forms were uniquely numbered and the responses recorded in a spreadsheet database. Questionnaires "returned to sender" uncompleted were recorded as undelivered.

For question 5, the value of the biopesticide projects where given were converted to US dollars where necessary using the internet currency converter provided by OANDA (http://www.oanda.com/). The same date of 1 September 1999, by when most researchers should have received the questionnaire, was used for all currency conversions. For questions 8 to 10, any responses encountered which were not clear (e.g. more than one box ticked instead of numbered to indicate priority) were recorded as a null response and omitted from the analysis if the answer could not be clarified using the biopesticide project title given as an answer to question 4. For questions 11 to 14, null responses were recorded if no answer or more than one answer for the same question was given. The only exception was question 12; if no answer was given, it was assumed that the respondent had no experience in this subject and it was recorded as "none".

Results of the Survey

Delivery and Response

During the period August to December 1998, 122 questionnaires were returned from respondents in 30 countries and 19 questionnaires were returned undelivered. It was assumed that 1019 questionnaires therefore reached their intended destination, giving an overall response rate of 12%.

Regional

Countries were grouped into geographical regions of Africa, Asia, Eastern Europe, Middle East and Latin America and the number of questionnaires sent to and the number of responses received from each region is shown in Figure 1. Response rates were highest for Latin America (16%) and the Middle East (17%), about the overall average for Asia (11%) and Eastern Europe (12%) and low for Africa (7%).

The greatest number of biopesticide researchers sent questionnaires were working in Asia. At 565, this was almost three times the next highest number of 191 researchers in Latin America. However, the response from Latin America was disproportionately higher; the number of 61 questionnaires returned from Asia was only about twice the number of 29 returned from Latin America.

The low number of responses from Africa is in contrast with the other regions; despite sending a reasonable number of 111 questionnaires to researchers in Africa, only nine replied.

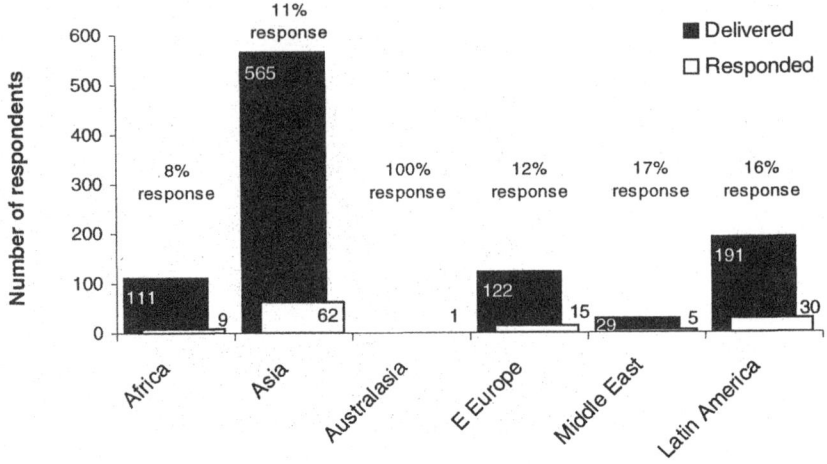

Figure 1. The number of questionnaires delivered and responses by region.

By Country

The number of questionnaires sent to and the number of responses received from researchers working in each country is shown in Figure 2. There was great variation in the number of questionnaires sent to each of the 50 developing countries which suggests the differences in research activity between them. Countries such as China and India are the most active with the largest number of biopesticide researchers (227 and 174 respectively). Other developing countries with relatively high numbers of researchers were Brazil (89), Russia (50), Egypt (47), Korea (46), Mexico (46) and Taiwan (41).

   The proportion of researchers responding to the questionnaire was variable between countries and appeared independent of the number of questionnaires sent to each. Among the countries with high numbers of researchers, response rates above the overall 11.8% were observed for India (14% replied), Brazil (13% replied) and Mexico (13% replied). However, despite sending out more than 40 questionnaires to each, the response rates were very low in comparison from China (6%), Russia (4%) and Egypt (4%). These differences might be explained by differences in the proportion of researchers with sufficient knowledge of the English language needed to understand and complete the questionnaire but this cannot be determined from the results.

   For the 42 countries sent less than 40 questionnaires, no replies were received from 20 countries of which the highest number were countries in Africa; eight out of 12 African countries sent a questionnaire did not reply. However, all of these unresponsive countries were sent only six or less questionnaires. At this low level of sample sizes, knowledge of the English

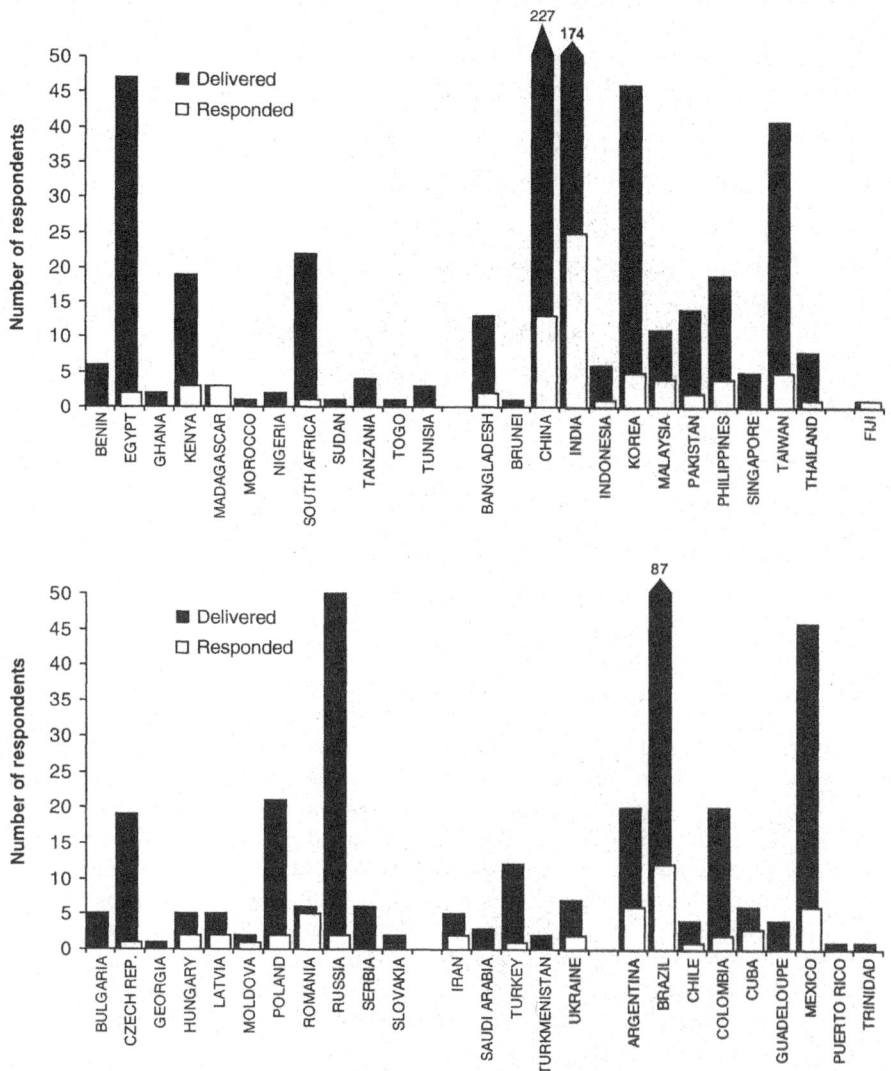

Figure 2. The number of questionnaires delivered and responses by country.

language and other possible factors affecting the response rates such as currency of contact details and efficiency of delivery in a particular country may become more apparent.

Type of Employment of Respondents

Of the questionnaires sent out, 95% were sent to researchers at academic establishments (universities or research institutes) and this is reflected in the returns where 93% of respondents worked for research institutes or universities (Table 1). Replies were also received from researchers working for commercial companies, plant protection services, an extension service and a regulatory authority.

Table 1. The types of employment of the recipients and respondents.

Type of employment	No. of recipients	Proportion (%)	No. of respondents	Proportion (%)
University	631	62	54	44
Research Institute	339	33	60	49
Commercial Company	14	1	4	3
Plant Protection Service			2	2
Extension Service			1	1
Regulatory Authority			1	1
Unidentified	35	3		

Research Funding of Respondents

102 researchers responded when asked about the value of their biopesticide research programmes. The answers ranged from US$360 to US$300,000. However, it was not always stated whether the figures given were per annum or for the duration of a programme, which may be more or less than a year. Despite this, Figure 3 does give an indication of the size and distribution of funding of research programmes in developing countries around the world. It was expected that smaller amounts of funding are easier to come by and therefore most research programmes would be small to medium with a few "mega" programmes being funded. The pattern in Figure 3 is very similar to that expected. The majority of programmes (67%) received low to medium levels of funding (up to US$25,000) while the number of programmes in categories above US$25,000 decreased as funding increased in value. A minority 11% of programmes received more than US$100,000.

Research programmes in Asia were observed to range across all the funding categories although the majority were in the under US$5000 category (45%) (Figure 4). Programmes in Latin America also ranged across all categories but a higher proportion of them were in the middle two categories for US$5000 to US$100,000 (77% as opposed to only 41% for Asia). Funding of the researchers surveyed in Africa ranged from under US$5000 to US$100,000. Programmes with the least funding were in the Middle East and Eastern Europe where no individual programme received more than US$25,000. However, the low sample

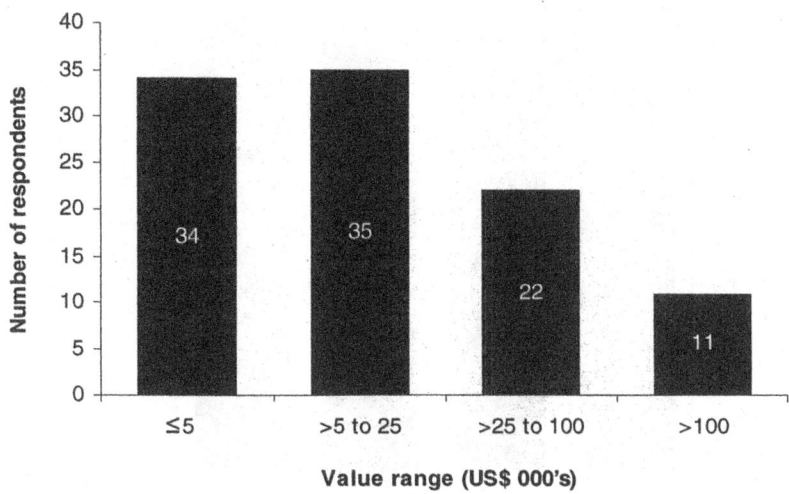

Figure 3. The value of biopesticide programmes among respondents.

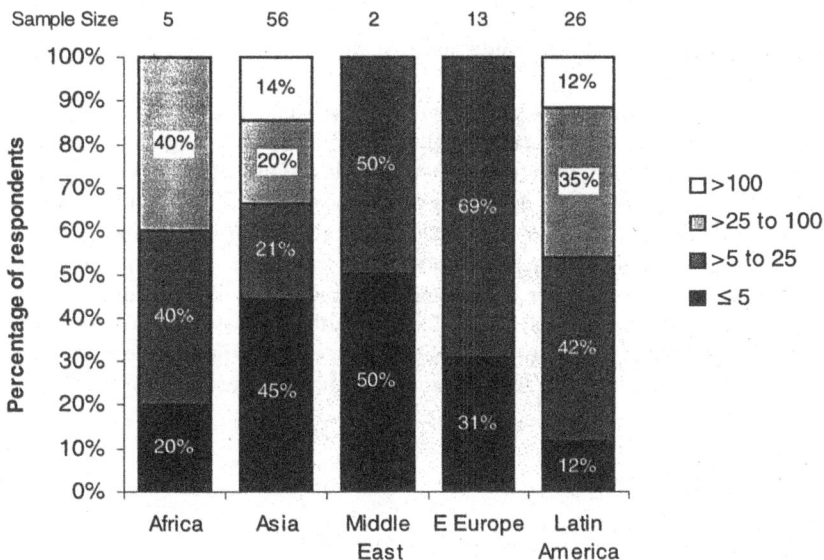

Figure 4. Proportional distribution of the funding received by respondents in each region.

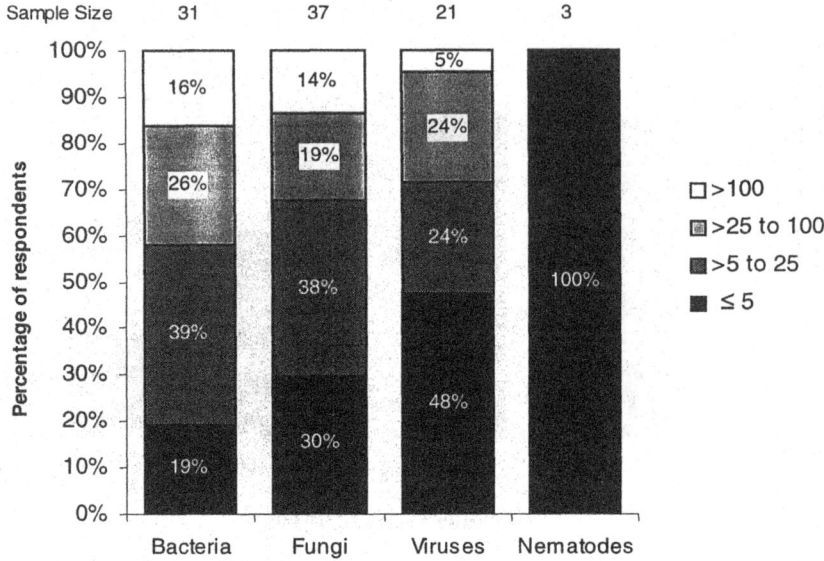

Figure 5. Proportional distribution of the funding received by respondents studying different control agents.

sizes for Africa, Middle East and Eastern Europe means it is debatable whether these observations are truly indicative of the actual levels of funding in these regions.

Research programmes working on bacteria, fungi and viruses were observed to range across all categories of funding (Figure 5). Higher proportions of research programmes whose primary interest is the use of bacteria as pest control agents receive funding in the categories above US$5000, with only 19% of programmes receiving under US$5000. The proportion of research programmes in the upper funding categories was less for those studying fungi, as the proportion receiving under US$5000 was higher at 30%. Funding of virus research programmes was even lower, with almost half of the research programmes (48%) in the under US$5000 category. Figures were available for only three nematode research programmes and all of these were under US$5000.

Research Interests

Researchers Listed in the Biopesticides Directory

Eighty-nine per cent of the researchers sent the questionnaire were listed in the directory of biopesticide researchers. Of these, 886 researchers were listed with their main research interests, allowing an analysis of the research interests of the majority of the researchers sent the questionnaire for comparison with those of the respondents.

Figure 6 shows that the greatest number of researchers were working on the control of insect pests (669 researchers). The number working on pathogens was much lower at 160, followed by 30 on weeds and 28 on nematode pests. In a plot of the relative proportions, there appeared to be no regional biases from the overall proportions towards working on one particular pest type (Figure 7).

Bacteria and fungi were the most popular types of control agents for research listed in the directory (343 and 328 researchers respectively; Figure 8). The numbers working on viruses and nematodes were lower at 107 and 81 researchers respectively. However, interest in investigating protozoa as control agents was extremely low, with only four researchers listed. The category antagonists contains only those research topics listed in the directory that did not specify the type of control agent studied; where an antagonist was specified, it was included in the appropriate category. It was assumed that unspecified antagonists would be either fungi or bacteria as no other type of antagonist was encountered in the analysis. The number of unspecified antagonists was also small in comparison to the fungi and bacteria categories and therefore the exclusion of unspecified antagonists was unlikely to affect the overall pattern.

An apparent regional bias was observed in the Middle East where, as a proportion of the total number of researchers for the region, relatively more researchers work on bacteria and protozoa as control agents and less on fungi than in other regions (Figure 9). Conversely, in Africa and Latin America, relatively more researchers work on fungi as control agents and less on bacteria than in other regions while in Eastern Europe, relatively more work on nematodes than in other regions.

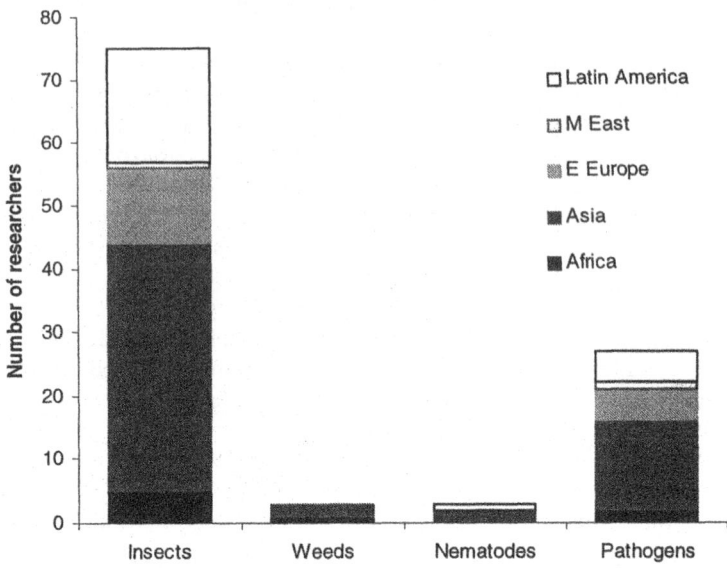

Figure 6. Pests studied by developing country researchers listed in the directory.

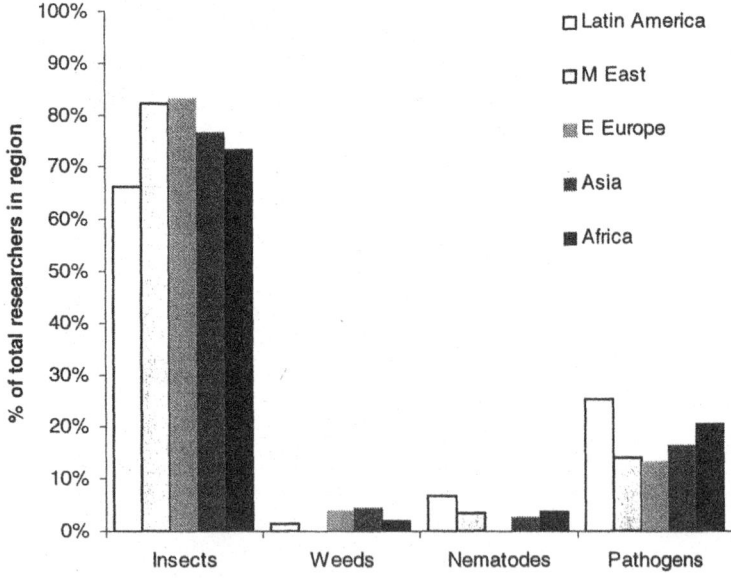

Figure 7. Proportions of developing country researchers listed in the directory studying different pest types.

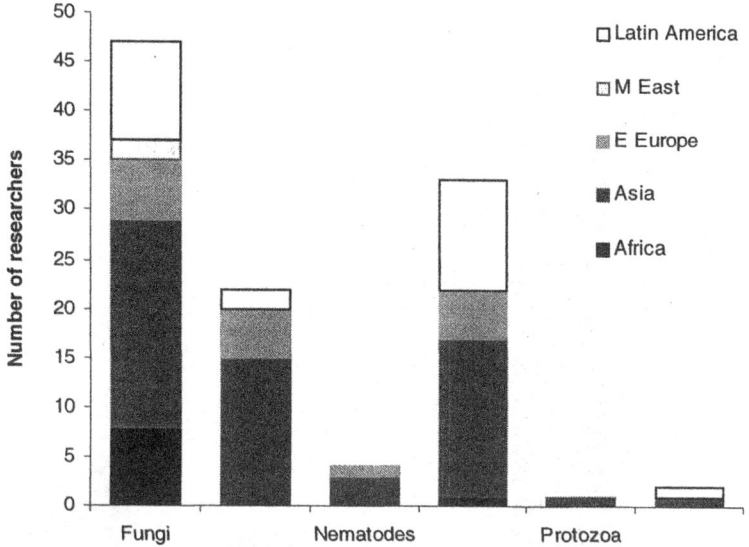

Figure 8. Control agents studied by developing country researchers listed in the directory.

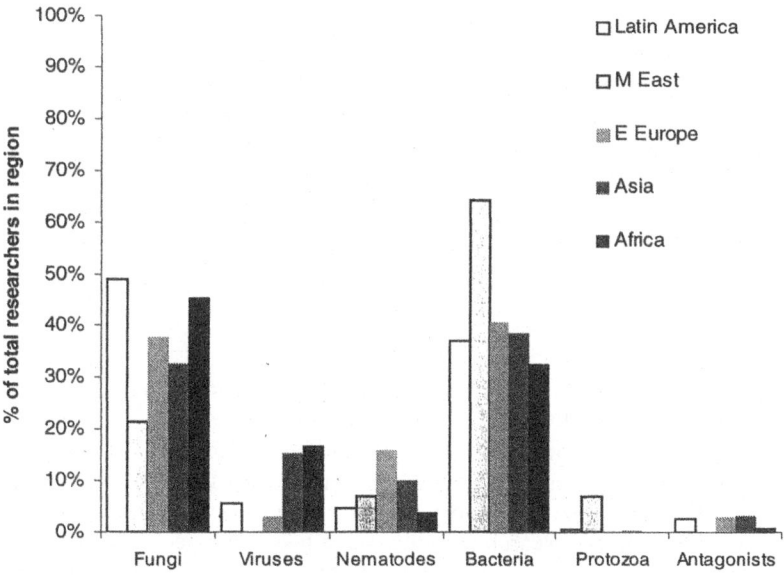

Figure 9. Proportions of developing country researchers listed in the directory studying different control agents.

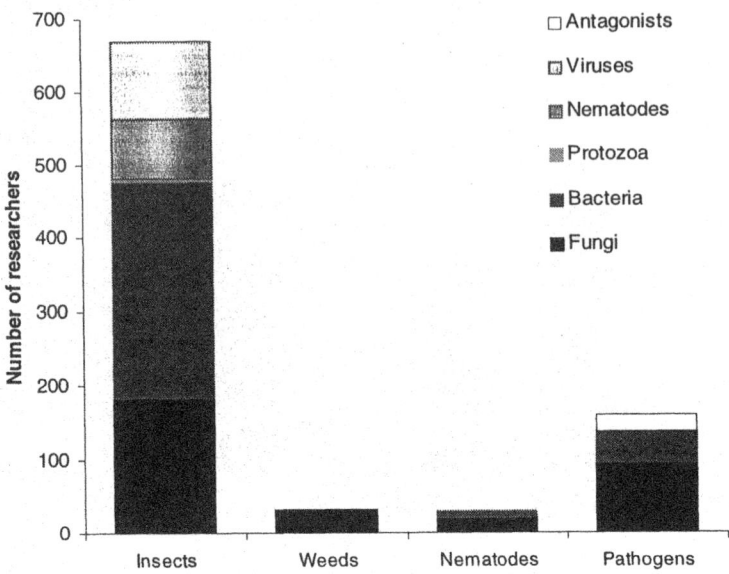

Figure 10. The agents studied for the control of different pest types by developing country researchers listed in the directory.

Bacteria, fungi, viruses and nematodes were all subjects of a number of researchers working on control agents for use against insects, bacterial agents being the most popular (Figure 10). Only a small minority of four researchers worked on protozoans and all were investigating their use to control insects. Similarly, viruses were only studied by researchers for the control of insects. Fungal biopesticides exclusively were being researched for the control of weeds and only fungal and bacterial biopesticides for the control of nematodes and pathogens.

Respondents

Among the respondents, 118 provided an acceptable answer to question 9 and of these, a majority of 69% listed the control of insect pests as their primary research activity and another 25% listed pathogens (Figure 11). Respondents' interest in other pests was very low. Weeds and nematode pests only scored three answers each and an extra category for ticks only one answer. However, if the secondary and tertiary research activities were taken into account, twice as many researchers were interested in nematode pests as in weeds. The pattern of these answers is almost identical to the pattern found in the directory shown in Figure 6.

The majority (69%) of the respondents to the questionnaire were listed in the directory, a drop in proportion compared to the questionnaires sent out (89%). This may reflect less interest in promoting biopesticide development generally among the large number of biopesticide researchers in the directory working on bacterial control agents of which there are already many examples of commercialised products. Figure 11 appears to support this; more of the respondents are working on fungi than bacteria, an opposite pattern to that shown in the analysis of the directory researchers (Figure 8). However, this difference might also be explained by the inclusion of respondents not in the directory but taken from delegate lists of conferences, where there may have been a higher proportion of researchers interested in fungal biopesticides. Unfortunately this could not be checked as the delegate lists did not include areas of research interest.

An acceptable answer to question 10 was provided by 110 respondents. Fungi were the most popular control agent for a primary research activity, followed by bacteria and then viruses (42%, 31% and 21% respectively; Figure 12). Interest in nematodes and protozoa was very low in comparison, although if numbers for secondary or tertiary research activity were included, up to 11% of respondents were interested in each of these categories. Only two respondents worked on plant products. Again, the pattern of these answers is similar to the pattern found in the directory in Figure 7.

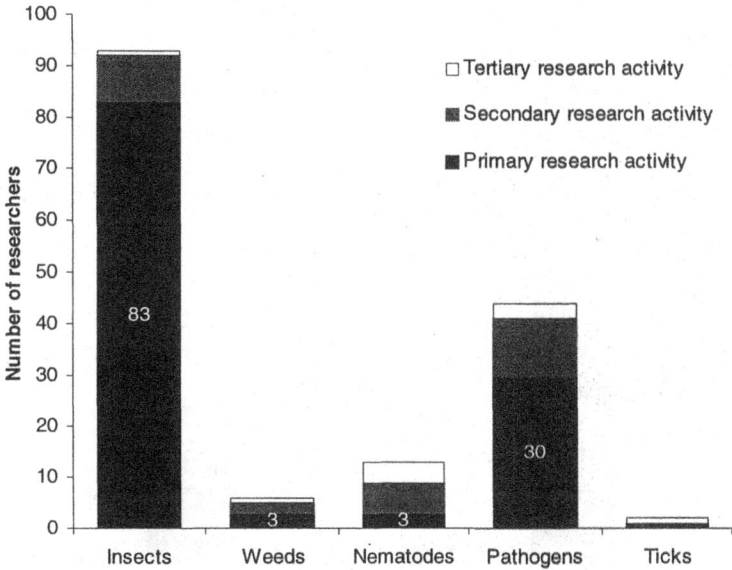

Figure 11. Types of pests confronted by the respondents (sample size = 118).

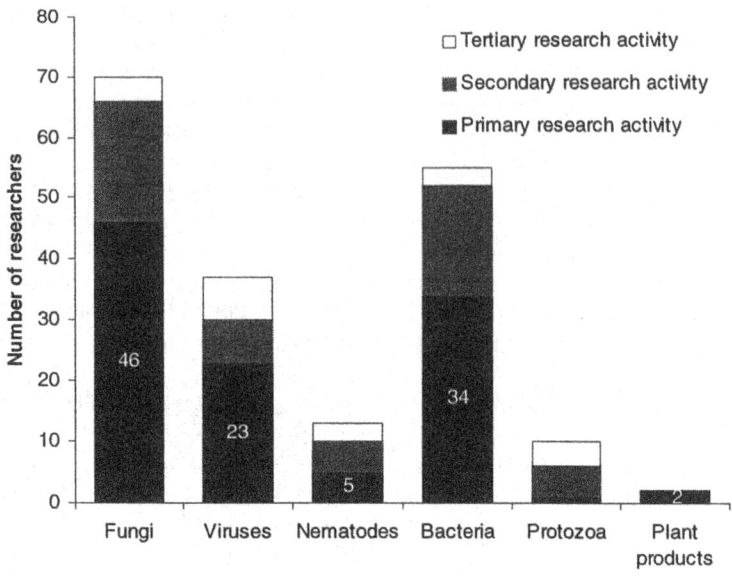

Figure 12. Types of control agents used by the respondents (sample size = 110).

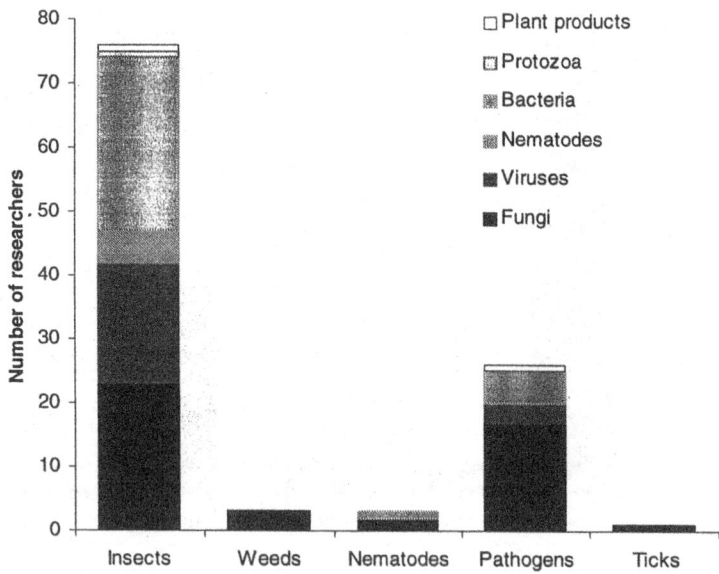

Figure 13. The types of control agents studied for the control of different pest types by the respondents (sample size = 109).

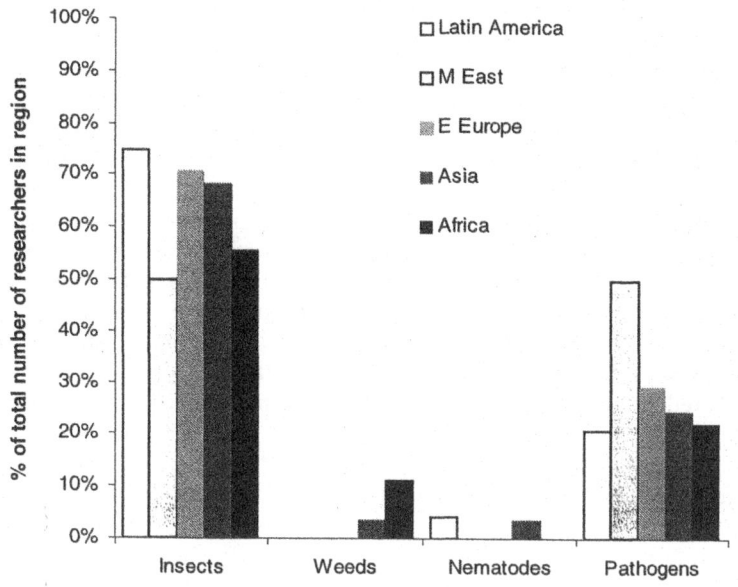

Figure 14. Proportions of respondents studying different pest types (sample size = 118).

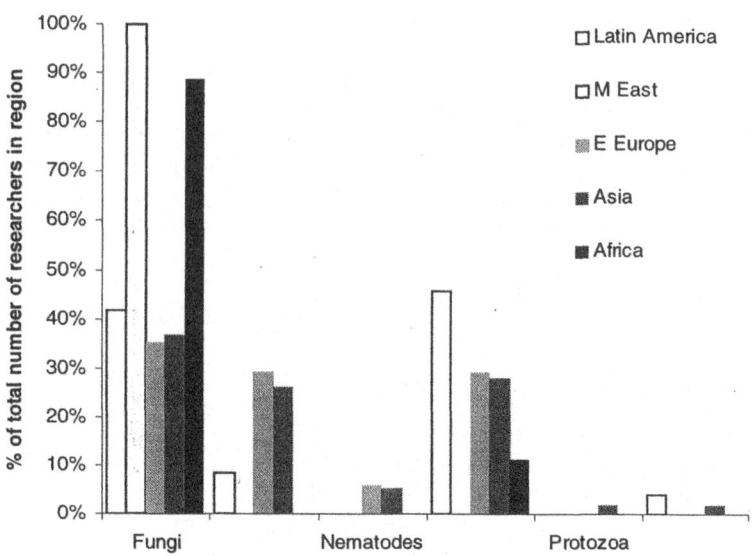

Figure 15. Proportions of respondents studying different control agents (sample size = 110).

The type of control agents used against each pest type by the respondents (Figure 13) also followed a very similar pattern to that seen in the directory analysis. A range of control agents were being researched for use against insect pests while only fungi were being used against weeds and fungi and bacteria against nematodes. The only exception was that some respondents were investigating viruses and protozoa for use against pathogens as well as fungi and bacteria.

The only apparent regional difference among the respondents in their replies to questions 9 and 10 was the low proportion in Latin America who were working on viruses in comparison to Asia and Eastern Europe (Figure 14).

Of the 107 respondents to question 8, the majority of 68% listed pest systems in agricultural crops as their primary research activity followed by 15% for horticultural crops, 5% for forestry and 5% for medical (Figure 15). Primary activity on stored products and urban and domestic pest systems was low but including secondary and tertiary figures, between 5 and 10% of respondents expressed some research activity. However, less than 5% of respondents undertook any research activity on each of livestock, veterinary or rangeland pest systems.

Constraints to Biopesticide Research and Development

Information

How much is a lack of access to information (international journals and books) a limitation to your project(s) in Biopesticide R&D?
Most researchers find that a lack of access to information is a significant limitation to their biopesticide R&D projects. On a sliding scale from 1 (Severe Limitation) to 5 (No Limitation), a majority of 61% of respondents answered between 1 and 3 (Figure 16). Only 16% said that a lack of access to information caused no limitation.

How would you define your own access to international and national journals and books relevant to biopesticides?
Most researchers find that their own access to information is limited. On a sliding scale from 1 (Severely Limited) to 5 (Unlimited), a majority of 71% of respondents answered between 1 and 3 (Figure 17). Only 8% experienced unlimited access to information.

Would there be value in summary information tailored to meet your needs for Biopesticide R&D?
There was a clear response to this question where all researchers thought that summary information would be of some value and a majority of 59% stated that it would be of high value (Figure 18).

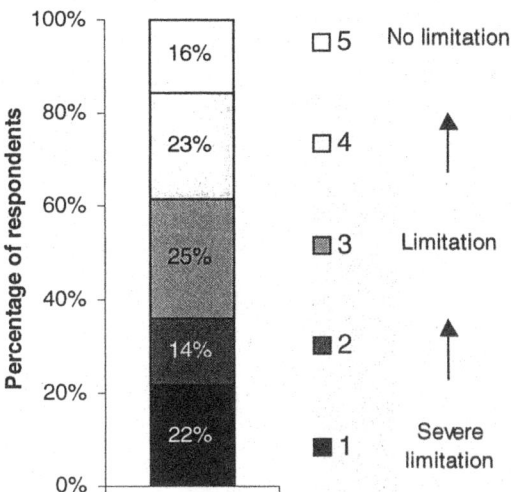

Figure 16. Availability of biopesticide information as a limitation to research (sample size = 114).

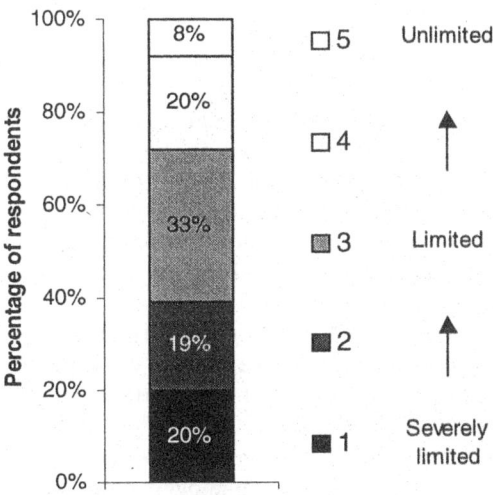

Figure 17. Respondents' own access to information on biopesticides (sample size = 115).

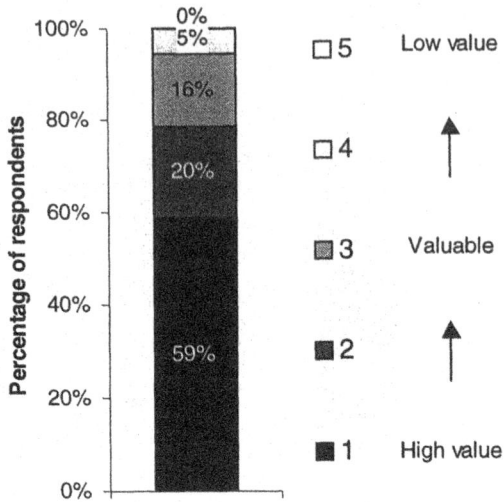

Figure 18. Value to respondents of summary information tailored to meet the needs of biopesticide R&D (sample size = 112).

Technical Expertise

How would you rate your expertise in each of the following specialist areas?
The responses to this question are shown in Figure 19. In this figure, the specialist areas have been arranged in a slightly different order to how they appeared on the questionnaire. Specialist areas considered to be associated with the initial stages of biopesticide development are on the left of the figure while those areas associated with the more advanced stages of development and commercialisation are to the right.

Figure 19 displays a trend of decreasing expertise among the respondents in specialist areas further down the line in the process of biopesticide development. If the areas are ranked in decreasing order of expertise, five of the top six areas of expertise appear to the left of the chart. Over 60% of the researchers surveyed felt that their level of expertise was good, leading or expert in pathogen exploration, pathogen identification, pathogen bioassay, research mass production and storage evaluation, all of which are the first steps in biopesticide development. There was a particularly high level of expertise among the respondents in pathogen exploration, pathogen bioassay and research mass production; over 15% declared themselves experts in each of these areas. The majority of researchers also thought their expertise in application technology to be good or better.

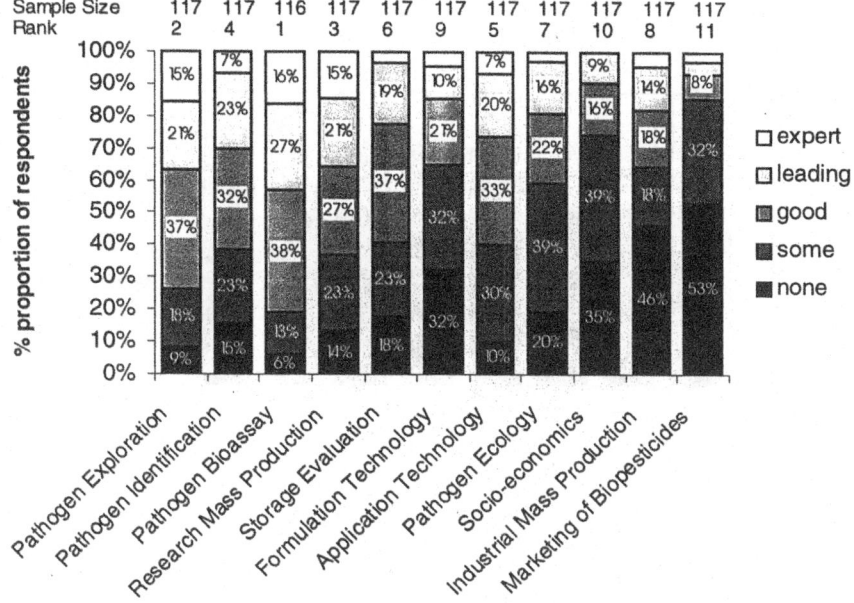

Figure 19. Respondents' ratings of their expertise in specialist areas of biopesticide development.

However, respondents expressed lower levels of expertise for the remaining areas associated with the more advanced stages of biopesticide development and commercialisation. Over 60% of researchers stated their level of expertise as some or none in formulation technology and industrial mass production. This increased to 74% for socio-economics and 85% for marketing. In addition, the proportions of those with no experience at all were particularly high in these four specialist areas. Expertise in pathogen ecology was also low, 59% responding with little or none.

Access to Specialist Equipment

Have you access for your own project(s) to the following specialist equipment?
The responses to this question are shown in Figure 20. In this figure, the equipment categories have been arranged in a slightly different order to how they appeared on the questionnaire. Relatively standard equipment for pathogen isolation such as a light microscope and autoclave is at the left of the figure while more specialised equipment is to the right. This is a comparable to the order of specialist areas used above; as researchers advance further in the process of biopesticide development and commercialisation, they will need more specialised equipment for identification of pathogens and to develop appropriate storage, formulation and application technologies.

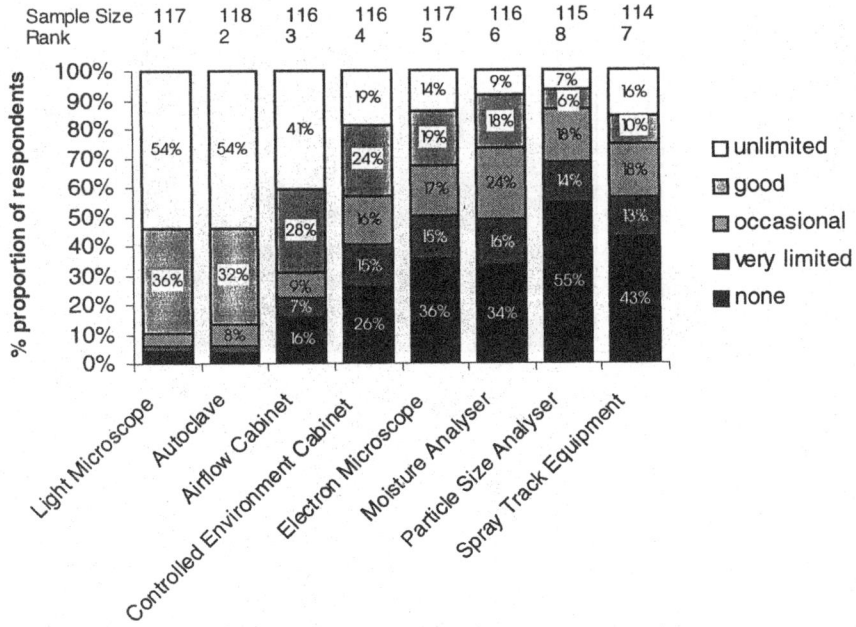

Figure 20. Respondents' access to specialist equipment.

There is a trend from left to right in Figure 20 of decreasing availability to researchers as equipment becomes more specialised. Access to light microscopes and autoclaves is good or unlimited for over 85% of researchers and for airflow cabinets it is 70%. Access to equipment seen as increasingly specialised is more restricted; about half of the researchers surveyed have very limited or less access to an electron microscope or moisture analyser and a third have no access at all. Access to equipment associated with application technologies is particularly poor with the majority of researchers having very limited or no access to a particle size analyser or spray track equipment.

Priorities in Research Funding

To achieve the successful development and commercial exploitation of biopesticides, which priority should be assigned to funding research in the following areas?
Over 85% of respondents felt that all the categories presented on the questionnaire should be medium to high priority for funding (Figure 21). If proportions of high and highest priority are summed and the categories ranked, the greatest proportion felt that priority should be given to funding research into formulation technology, closely followed by application technology, field trials

and industrial mass production. The pathogen exploration, research mass production, pathogen identification and pathogen bioassay categories were all similar in attracting a slightly lower proportion of respondents who thought they were a priority for research funding. The lowest proportions of researchers felt that ecotoxicology, storage evaluation, pathogen ecology, economic analysis of demand and economic analysis of supply were a priority for funding.

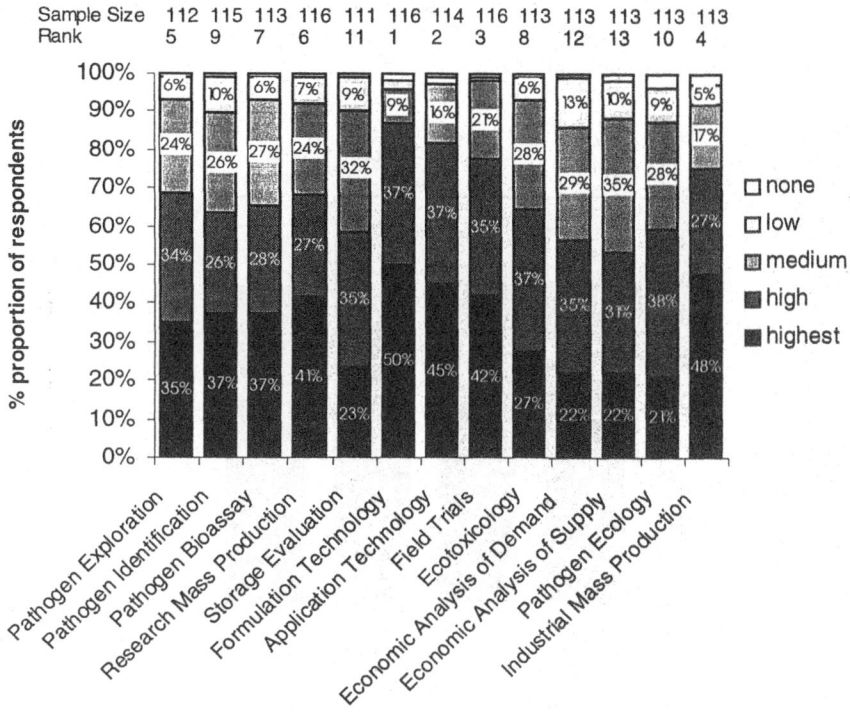

Figure 21. Respondents' priorities for funding in biopesticide research.

Barriers to Commercialisation

Know-how

Please evaluate the limitations to the potential commercialisation of biopesticides from your own project(s) in the following areas.

Figure 22 shows the proportional response for each of the know-how categories in this question, all of which are a potential limitation to the commercialisation of biopesticides. Over 75% of all respondents thought each category was a moderate to very severe limitation in their biopesticide project(s).

At 69%, the largest majority of respondents thought that problems with scale-up were a severe or very severe limitation to commercialisation. A further 23% thought it was a moderate limitation and only 8% of respondents thought it was either a small or no limitation.

For the other three know-how categories, respondents thought the limitations were lower but not greatly so. A lack of business skills, of understanding of commercialisation or of economic analysis were all thought to be a severe or very severe limitation to commercialisation by over 50% of respondents. In the minority, 6 to 8% of respondents thought that they were no limitation and 12 to 15% thought they were a small limitation.

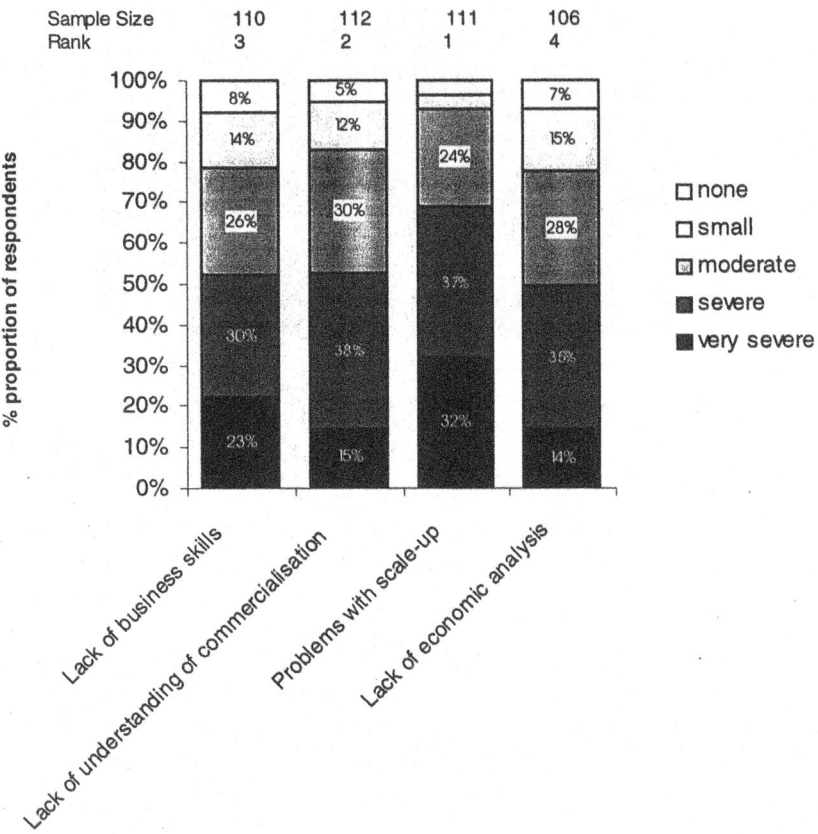

Figure 22. Respondents' evaluation of the know-how barriers to the commercialisation of biopesticides.

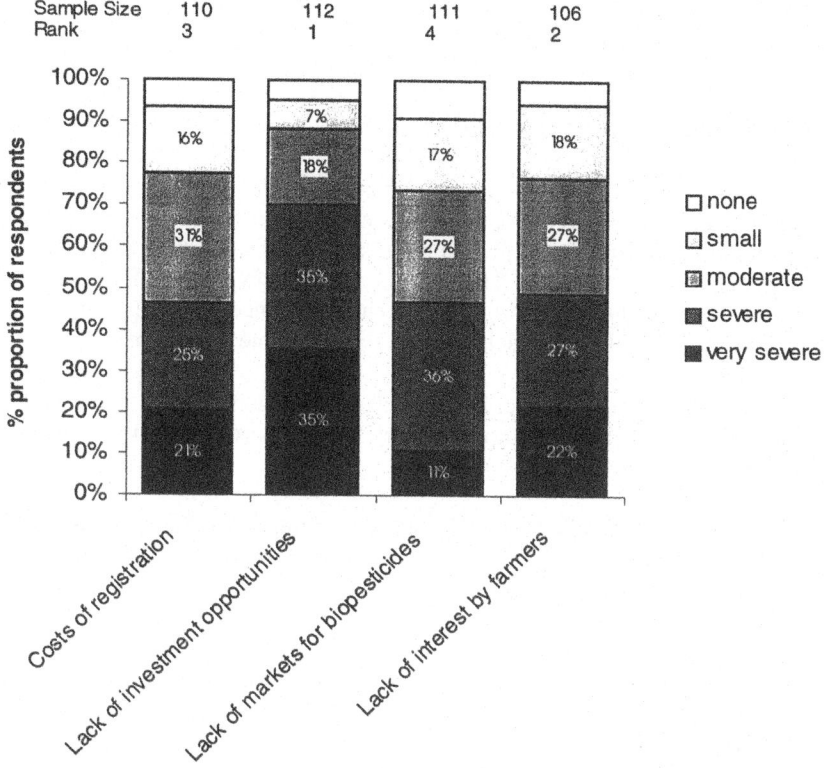

Figure 23. Respondents' evaluation of the financial barriers to the commercialisation of biopesticides.

Financial and Markets

Please evaluate the limitations to potential commercialisation of biopesticides from your own project(s) in the following areas.
Figure 23 also shows the proportional response for each of the financial and market categories in this question, all of which are also a potential limitation to the commercialisation of biopesticides. Over 70% of all respondents thought each category was a moderate to very severe limitation in their biopesticide project(s). Respondents who thought these categories were small or no limitation were all in the minority, from 12 to 27%.

A lack of investment opportunities was perceived by the highest proportion of respondents (70%) to be a severe or very severe limitation to commercialisation (Figure 23). The costs of registration, lack of markets and lack of interest by farmers were thought to be severe or very severe limitations by a lower but still sizable proportion of between 45 and 50% of respondents.

However, of these, the lowest proportion of respondents perceived a lack of markets for biopesticides to be a very severe limitation.

Differences Between Specialisms

Differences in the answers given to questions 11 to 14 were looked for also between groups of researchers principally working on fungal, bacterial and viral biopesticides. Protozoa and nematode researchers were not included as the sample sizes were too small. The responses are shown in Figures 24 to 31 in Appendix 1. Mostly there were no great differences to the overall trends observed above for all respondents but there were some exceptions:

- Overall, the viral biopesticide researchers indicated slightly more expertise, greater availability of equipment and lower research funding priorities in comparison to the other two groups.

- The expertise in application technology was thought to be good to expert by 80% of the viral group, higher than the other two groups (50% for fungi and 60% for bacteria).

- 55% of viral biopesticide researchers had occasional to unlimited access to a particle size analyser, higher than the other two researcher groups for bacteria and fungi (about 25% in both; also about 60% in both these other two groups had no access at all).

- There were some differences in the priorities in research funding between the three groups of researchers although all felt that formulation and application technology were the highest priority for funding. Pathogen bioassay, economic analysis of demand and supply were ranked higher and research mass production and ecotoxicology ranked lower by bacterial biopesticide researchers than the viral and fungal groups. Pathogen exploration, ecotoxicology and pathogen ecology were ranked higher and field trials lower by viral biopesticide researchers than the other two groups. Fungal biopesticide researchers ranked research mass production higher compared to the other two researcher groups.

- A lower proportion of the bacterial biopesticide researchers thought that problems with scale up were a severe or very severe limitation to biopesticide commercialisation (57% compared with 80% for the viral group and 73% for the fungal group) although all three groups ranked it as the greatest limitation. Fungal and bacterial biopesticide researchers consider a lack of economic analysis as more of a limitation to biopesticide commercialisation (ranking it 2nd) than the viral group who ranked it the least limiting out of the four categories.

- Out of the four categories, costs of registration was ranked as the least limiting to biopesticide commercialisation by the bacterial biopesticide researchers, of whom only 17% thought it was a severe or very severe limitation. This is in contrast with the fungal and viral groups who ranked it second and over 50% in these two groups thought it was a severe or very severe limitation.

Differences Between Regions

Differences in the answers given to questions 11 to 14 were looked for between groups of researchers working in Asia, in Latin America (two regions with a reasonable sample size of more than 20 respondents) and in the "other" remaining areas of Africa, Eastern Europe and Middle East grouped together. The responses are shown in Figures 32 to 39 in Appendix 2. Mostly there were no differences to the general trends observed above for all respondents but there were some exceptions:

- In Latin America, 64% of respondents thought that access to information was either not limiting or a slight limitation to their biopesticide projects. In Asia and the "other" regional group, a much lower 31% and 26% gave the same response. Also, no-one considered it a severe limitation in Latin America whereas 31% in Asia and 22% in the "other" regional group thought it was.

- A similar response was also seen for the next question where researchers in Latin America considered their own access to information higher than did researchers in other regions.

- There were some noticeable differences in the priorities in research funding between the three regional groups of researchers. Biopesticide researchers in Latin America ranked ecotoxicology higher and storage evaluation and ranked industrial mass production lower compared to the other two regional groups. Biopesticide researchers in Other Countries ranked ecotoxicology lower compared to the other two groups. Pathogen bioassay and research mass production were ranked higher and pathogen identification and pathogen ecology lower by biopesticide researchers in Asia than the other two groups.

- Problems with scale-up were felt to be more limiting by a higher proportion of respondents from Asia and the Others regional group than lack of business skills, of understanding of commercialisation or of economic analysis. This peak was not observed amongst Latin American respondents who in terms of proportions rated all four know-how categories similarly.

- Lack of markets for biopesticides ranked last as a severe or very severe limitation in the Others regional group and a higher proportion of respondents thought that costs of registration was a severe or very severe limitation and this category ranked second. This is an opposite response to Asia and Latin America where lack of markets for biopesticides was second and costs of registration was 4th and there was also a smaller difference in the proportional response between the two categories.

Discussion

In compiling an extensive but non-exhaustive list of over 1000 biopesticide researchers from 50 developing countries world-wide, it was evident that there is a substantial level of interest and availability of experience in the development of biopesticides. However, the very low availability of successful biopesticide products on the market in developing countries does not reflect their activities. This survey was the first to be conducted with the objective of gathering evidence of the main constraints to biopesticide development experienced by researchers in developing countries. The response rate of 12% from 30 countries was excellent for an unsolicited mail-shot questionnaire of this type and is further evidence for the desire in developing countries to develop effective biopesticide products.

The list of recipients of the questionnaire was considered to be representative of biopesticide research as a whole in developing countries because it included a large sample of researchers world-wide working on all the major areas of biopesticide research. The sub-sample of 122 respondents was in turn considered to be a representative sample; there was a world-wide geographic distribution of respondents and the research areas worked on and the type of employment of researchers followed similar patterns between the recipients and respondents. The majority of biopesticide researchers were based at publicly funded research institutes and universities and R&D of fungal, bacterial and viral control agents for the control of insects in agricultural and horticultural crops is dominant. Research on other pest systems and nematode and protozoan agents was being undertaken but at much lower levels.

The results of the survey clearly indicated a number of factors constraining biopesticide development in developing countries. The areas of expertise listed in the questionnaire were based on the steps necessary for biopesticide development as outlined by Dent (1997) and for the purpose of analysis were ordered in developmental sequence. The expertise among researchers was found to be high in the areas associated with the initial stages of the biopesticide development process (exploration, identification, bioassay and research mass production) but lacking in the areas associated with the later developmental steps (e.g. formulation, ecology, socio-economics, registration and commercialisation). A similar trend was also seen for the availability of facilities, where equipment associated with the later stages of development was less accessible to researchers.

Appropriate formulations of biopesticides can improve product stability, viability and may reduce inconsistency in field performance. Ecological studies of host-pathogen interactions such as secondary cycling of the biocontrol agent (Thomas *et al.*, 1997) provide valuable information that can be used to define "good use strategy". Socio-economic studies are necessary to identify all costs of scale up and production and whether the demand exists for the biopesticide to ensure that an effective product can be economically produced. Registration requires the collation of data and preparation of a dossier for submission to a national regulatory authority which is a specialised skill different from that for scientific documents. In addition to toxicological and ecotoxicological and product performance issues, registration also addresses matters such as product labelling, disposal of the active ingredient and the formulated material. The commercialisation process involves identification of appropriate companies (based on their manufacturing capability, product range, access to capital and markets, sales and distribution systems, size, location and flexibility), negotiation over IPR issues (confidentiality agreements, license agreements), transfer of technology (production scale up trials), product registration and establishing suitable quality control standards and procedures (Dent and Lomer, 1999). Without these latter areas of expertise, it is easy to see why so few biopesticide products are successfully developed.

Many researchers considered themselves as having expertise in application technology but it is arguable that this is an overestimation on their part. An analysis by Dent (1997) of the amount of research undertaken showed that application technology for biopesticides is relatively underdeveloped and well below that for chemical insecticides, indicating a general world-wide malaise in application research. However, that it is essential is not in doubt. Experience has shown that attention to the detail of biopesticide application technology can yield significant improvements in performance (Bateman, 1999). The key requirement is to ensure that the biopesticide is sprayed in such a way that the droplet size is optimised for pick-up by the pest and that the amount of biological agent within each droplet is sufficient to ensure infection and subsequent mortality. However, in order to achieve this, application technologists consider formulation properties in the spray tank, atomisation, transport to and impaction on the target surfaces, distribution of the deposit and subsequent environmental degradation, to biological effects (Chapple and Bateman, 1998; Bateman, 1999). Spray application technologists utilise the principles of physics and mathematics and experimental equipment such as spray tracks and particle size analysers to optimise dose transfer and mortality from a biopesticide. Thus, expertise in formulation and availability of such equipment is also necessary for expertise in application. However, this formulation expertise and equipment was generally not available to the respondents and brings the higher level of expertise expressed in application technology into doubt.

The funding priorities of the respondents demonstrated an overall acceptance that all research areas listed were stages necessary for biopesticide development. It is interesting to note however, that these priorities did not always reflect the levels of expertise among respondents; it might be expected that areas lacking available expertise would have a higher funding priority in order to improve on the deficiencies identified. There was some variation in the degree of prioritisation between the research areas and this may reflect the perceived need of the respondents at the time of the survey. Formulation, application technology and field trials were ranked highly and may have been where most respondents' development projects were being limited at the time. Once these have been addressed, they will encounter new problems as they progress possibly associated with storage, ecotoxicology, pathogen ecology and economic analysis, which were given lower funding priority in the survey, and their priorities for funding research into these will increase.

That these particular areas were ranked amongst the lowest priorities indicates that they are viewed as less important areas that can be left until the later stages. They are, however, as important as other areas as a source of problems, which can be the eventual failure of a biopesticide under development. Biopesticide products usually need to have a minimum shelf life of one year but poor storage capability quickly results in loss of viability and potency, growth of contaminants and loss of formulation properties (Dent, 1999). Pathogen isolates must be assessed to ensure that they have a minimal impact on the environment and organisms other than the target pests. A lack of knowledge about the host-pathogen interactions results in poor use strategies, which decreases the effectiveness of a biopesticide and biopesticide products will not be sustainable if they cannot be economically produced. These areas could be considered far earlier in the developmental process in parallel with other research rather than as part of a stepwise process. Problems in these areas encountered at an early stage allow the direction of the research being undertaken to be reconsidered to address them; consequently there is a higher probability of developing successful solutions, thereby minimising the risk of failure and wastage of investment.

There are also likely to be some small biases in these priorities through favouritisms held by some respondents for their own specialisms. The initial stages of development ranked highly in funding priority among respondents despite the apparent proficiency and activity that already existed amongst researchers in these areas. While the discovery of potential agents is necessary in order to develop biopesticides, there may have been a bias caused by a concern that ranking these areas as a lower priority could threaten their current activities.

Availability of information is also constraint for the majority of biopesticide researchers. Over 60% of respondents found that a lack of access to information on biopesticides was a medium to severe limitation to their biopesticide research.

As discussed above, the expertise of respondents did not range equally across all the necessary stages of biopesticide development and this suggests that there is a need for information in order to undertake the later stages of development to develop biopesticide products. Indeed, 60% of respondents said that summary information tailored to meet their needs in biopesticide R&D would be of high value and 95% said medium to high value. Therefore, those respondents who considered lack of information to be not very limiting to their projects are probably conducting research at the initial developmental stages, where experience is already high, without progressing into new areas of expertise. Over 70% of respondents found that their own access to biopesticide information in international journals and books is limited to severely limited and as a consequence, these researchers will find it difficult to monitor the latest advances and increase their expertise.

Lack of business skills, lack of understanding of commercialisation, problems with scale up from research to industrial mass production and lack of economic analysis were all considered to be moderate to severe know-how constraints to commercialisation of biopesticide products by over 75% of respondents. That all these factors were perceived as limitations is indicative of a level of inexperience with the process of commercialisation described above among the respondents. Expertise in these areas is present as standard in the commercial sector and essential to the future development of biopesticides. This survey is further confirmation that it is rarely available in public research institutes or among the scientific establishment.

It was soon observed from the responses that the amounts of investment in biopesticide research and development was very low in developing countries. The majority of respondents were funded on small to medium programmes up to US$25,000 with fewer large-scale programmes of US$100,000 or more. Over 90% of researchers are based at publicly funded institutions and it would appear therefore that most of this investment is from the public sector. The large proportion of respondents with project funds below US$1000 indicates the existence of a number of available researchers with an interest in biopesticide research that is not being utilised. The low level, mostly public funding of the majority of these researchers did not appear sufficient for them to proceed all the way through the developmental stages. This was supported by the responses to the question on financial and market barriers to commercialisation, where lack of investment opportunities was perceived to be the greatest constraint by over 85% of respondents.

A lack of investment reflects a general lack of awareness of the potential value of biopesticides. There has been a universal failure of adequate promotion which has led to a lack of structural and financial support provided by policymakers (Waage, 1997). Similarly, awareness of commercial opportunities offered by biopesticides is lacking amongst commercial investors while a lack of promotion to farmers results in a lack of interest and demand for biopesticide products. Increased promotion of biopesticides amongst international and national agencies, decision makers, extension agencies, entrepreneurs, potential commercial investors and farmers is needed to encourage more demand for, and consequently more investment in, the development of biopesticide products.

The costs of registration, lack of interest by farmers and lack of markets were also considered financial constraints by respondents but to a lesser degree because overcoming these factors may have been seen as being dependent on a degree of investment. However, it often only requires relatively small amounts of investment to overcome these constraints and promotion of subsequent successes would then stimulate a wider investment in biopesticides.

Although generally cheaper than their chemical counterparts, biopesticide registration still involves costly toxicological studies and there is a lack of standardisation in registration procedures and regulation in developing countries (Newton *et al.*, 1996). This has meant that some studies have to be tailored differently to the needs of each country, creating additional costs. Assistance with the introduction and adoption of appropriate registration procedures in developing countries with a degree of standardisation would allow companies to produce one set of studies acceptable in more than one country and thus promote the wider uptake of biopesticides.

It is essential that farmer interest is developed and maintained to ensure a demand for biopesticides, without which there are no markets. The nature and mode of action of biopesticides needs to be explained to farmers who are used to chemical insecticides which are often fast acting and are visibly effective. Farmers are often reticent about adopting biopesticides that may take some days or even a week to kill the pest even though effective control is achieved through reduced feeding. Hence efficacy and cost-effectiveness needs to be proved in the field. In the developing country context, this training and technology transfer is likely to require the investment of public funds, at least in the initial stages; these funds need to be available and well targeted (Jones *et al.*, 1993).

Respondents considered a lack of markets the least constraining financial factor, indicating that there is a belief among respondents that markets for biopesticides exist. However, that it was considered a moderate to severe constraint by over 45% of respondents also indicates that many researchers are working blindly on developing biopesticides without knowledge of either the potential biopesticide markets or whether the biopesticide they are developing satisfies the product characteristics demanded by these markets. This is a consequence of the low levels of access to information and expertise in the economic analysis and marketing of biopesticides available to public research organisations.

The above discussion has concentrated on the overall picture that emerges from the results of the survey. However, small differences were observed between the responses of researchers grouped by either region or research specialism (in this case, type of control agent studied). The existence of these differences is not surprising; researchers of different specialisms will encounter different technical problems that arise from the differing biology of the control agents being studied. For example, bacteria can be produced in large quantities on a simple medium at a low cost (Smits, 1997) whereas the production of entomopathogenic fungi has not been so simple, requiring the development of solid state fermentation processes (Jenkins *et al.*, 1998), which are presently more costly and require large amounts of substrate to produce large quantities. This may account for the observations that there was a lower funding priority for research mass production amongst bacterial researchers than fungal researchers and that fewer bacterial researchers felt that problems with scale-up were a severe limitation to biopesticide commercialisation. Differences in the constraints experienced by researchers between regions are also expected on the basis that differences exist between countries in the type and degree of pest problems that they experience, their government policies towards the development and regulation of biopesticides and the demand for biopesticides as another option in pest management strategies.

The small sample sizes of some of these groups means that these observed differences between them can only be used as an indication of the variation in the constraints experienced by researchers and not confirmation. However, the existence of such variation which may relate to a researcher's geographical location and specialism, needs to be recognised when assessing the constraints to the successful development of biopesticide products and subsequent priorities for providing researchers with technical assistance and funding to overcome them.

Conclusions

These analyses confirm that the difficulties and constraints that are facing researchers in developing countries involved in biopesticides relate to lack of expertise in the crucial later stages of development and that the inappropriate model of biopesticide development described by Dent (1997) is being pursued in developing countries. Biopesticide research is receiving mostly low investment, largely from the public sector and rarely involves the multi-disciplinary expertise required to develop a biopesticide from start to finish. The successes of the LUBILOSA programme in developing and commercialising a biopesticide are attributed to the multi-national, multi-institute, multi-disciplinary approach which brought together the whole range of necessary expertise from a number of different institutes in collaborative programmes of research and development (Dent, 1999). The lessons learnt from this experience suggest that significant enhancements to future biopesticide research and development in developing countries can be made through the adoption of this alternative model. Targeted assistance in the form of specialised support, facilities and access to expertise on a multi-national, multi-institute, multi-disciplinary basis is required in developing countries in order to remove the constraints to the future successful development and utilisation of biopesticides.

References

Bateman, R.P. (1999) Delivery systems and protocols for biopesticides. In *Methods in Biotechnology*, Vol. 5, *Biopesticides: Use and Delivery* (eds Hall, F.R., Menn, J.J.), pp. 509–258. Humana Press, Totowa, New Jersey.

Chapple, A.C. and Bateman, R.P. (1997) Application systems for microbial pesticides: necessity not novelty. In *Microbial Insecticides: Novelty or Necessity?* British Crop Protection Council Proceedings/Monograph Series No. 68, 181–190.

Copping, L. (1998) *The Biopesticides Manual*, 333 pp. British Crop Protection Council, Farnham Royal, UK.

Dent, D.R. (1997) Integrated pest management and microbial insecticides. In *Microbial Insecticides: Novelty or Necessity?* British Crop Protection Council Proceedings/Monograph Series No. 68, 127–138.

Dent, D.R. and Lomer, C.J. (1999) Product development and commercialisation in development assistance projects: a case study - LUBILOSA (unpublished report). 25 pp. CABI/IITA.

Harris, J. (2000) *Chemical Pesticide Markets, Health Risks and Residues*. CAB International, Wallingford, UK.

Jeyaratnam, J. (1990) *World Health Statistics Quarterly* No. 43. World Health Organisation.

Jones, K.A., Westby, A., Reilly, P.J.A. and Jeger, M.J. (1993) Ch 14 in *Exploitation of Micro-organisms* (ed. Jones, D.G.). Chapman and Hall, London, pp.343–370.

Lisansky, S. (1997) Microbial biopesticides. In *Microbial Insecticides: Novelty or Necessity?* British Crop Protection Council Proceedings/Monograph Series No. 68, 3–10.

Newton, P.J., Neale, M.C., Arslan Bir, M., Brandl, M., Fidgett, M.J. and Greatrex, R.M. (1996) Full-range pest management with IPM systems —

an industry view of the options for non-indigenous biopesticides. *Biological Control Introductions — Opportunities for Improved Crop Production.* Proceedings of an International Symposium, British Crop Protection Council, Brighton, UK, 18 November 1996, 77–97.

Quinlan, R. (1995) *Biopesticides: World-wide Directory of Research and Researchers.* CPL Press, Newbury, UK.

Thomas, M.B., Blanford, S. and Lomer, C.J. (1997) Reduction of feeding by the variegated grasshopper, *Zonocerus variegatus*, following infection by the fungal pathogen, *Metarhizium flavoviride. Biocontrol Science and Technology* 7, 327–334.

Waage, J.K. (1997) Biopesticides at the crossroads: IPM products or chemical clones? In *Microbial Insecticides: Novelty or Necessity?* British Crop Protection Council Proceedings/Monograph Series No. 68, 11–19.

Waage, J.K. (1999) Beyond the realm of conventional biological control: harnessing bioresources and developing biologically-based technologies for sustainable pest management. In *Biological Control in the Tropics*, Proceedings of the Symposium on Biological Control in the Tropics, held at MARDI Training Centre, Serdang, Malaysia, 18–19 March 1999, pp. 5–17. CAB International, Wallingford, UK.

Appendix 1: Specialisms

Information

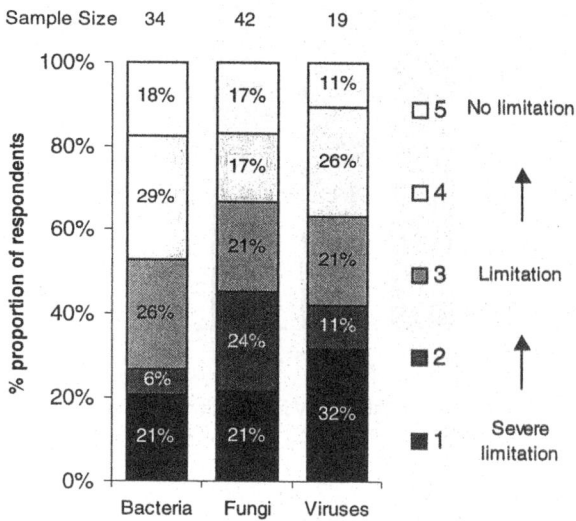

Figure 24. Availability of biopesticide information as a limitation to research.

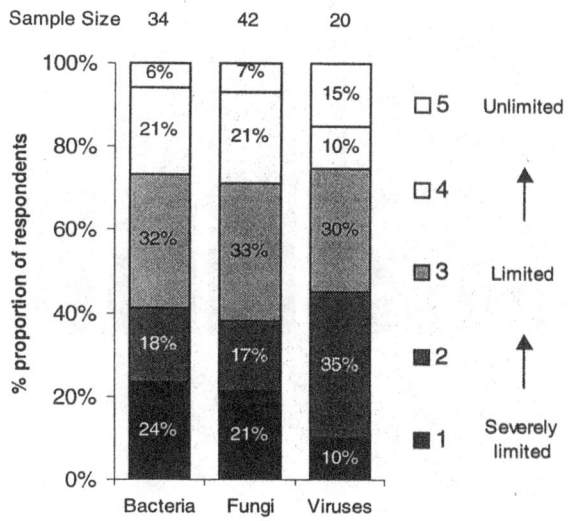

Figure 25. Respondents' own access to information on biopesticides.

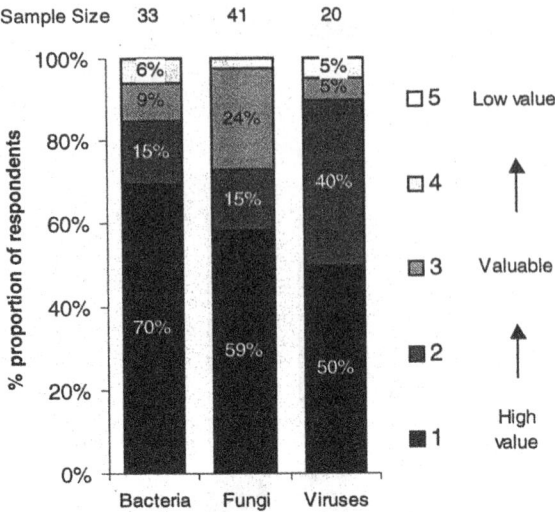

Figure 26. Value to respondents of summary information tailored to meet the needs of biopesticide R&D.

Technical expertise

a) Bacteria

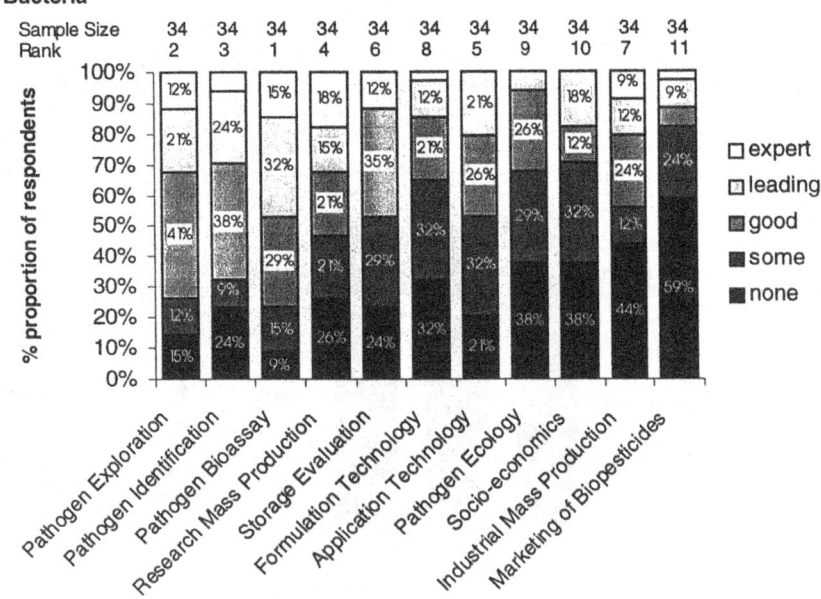

Figure 27. Respondents' ratings of their expertise in specialist areas of biopesticide development. *Cont'd.*

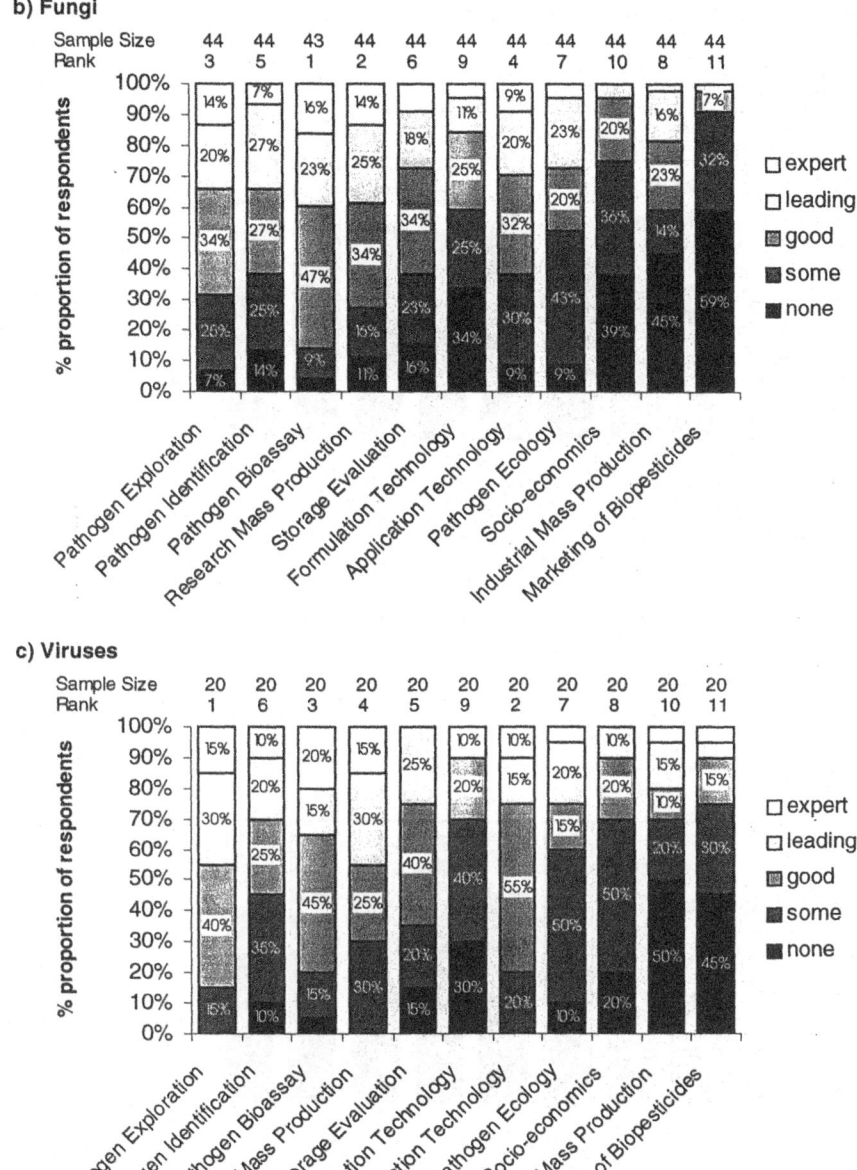

Figure 27, cont'd.

Access to Specialist Equipment

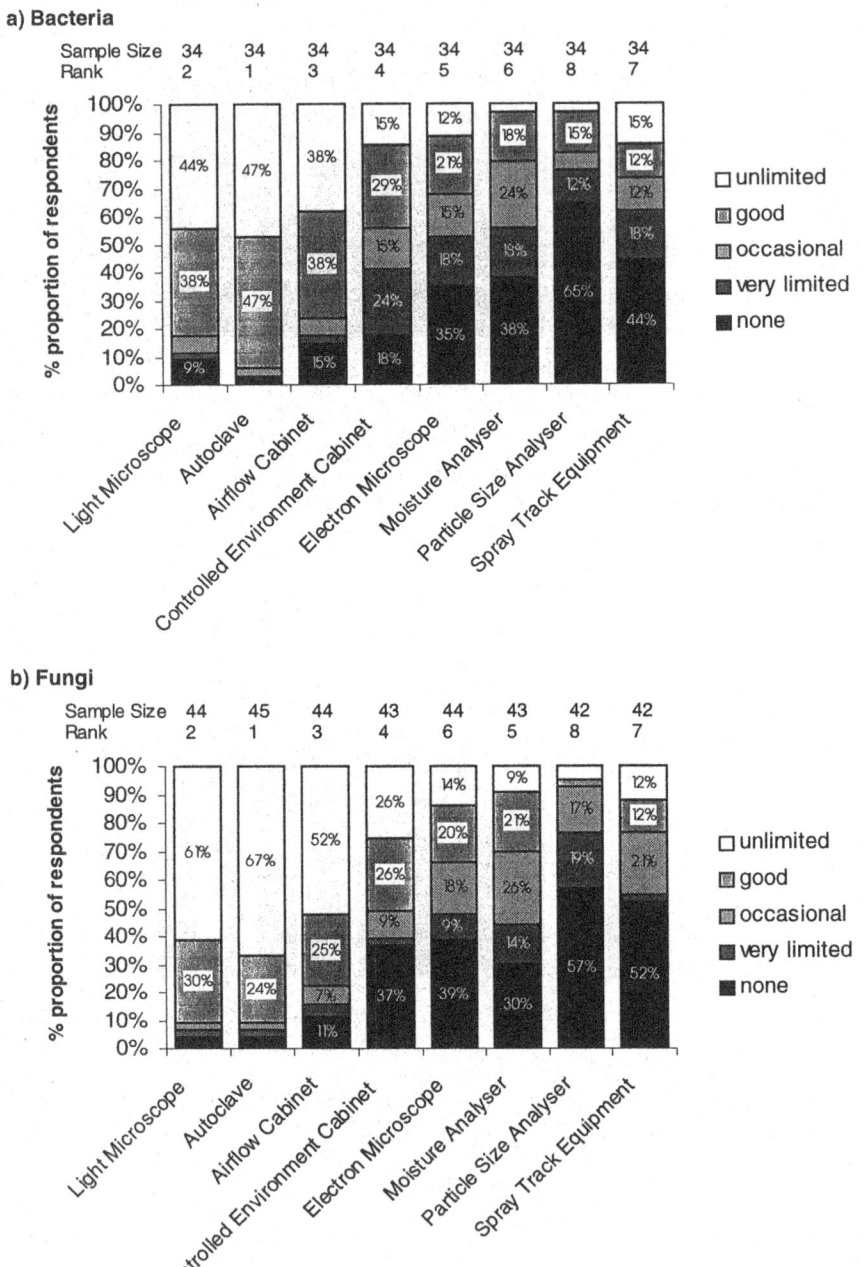

Figure 28. Respondents' access to specialist equipment. *Cont'd.*

c) Viruses

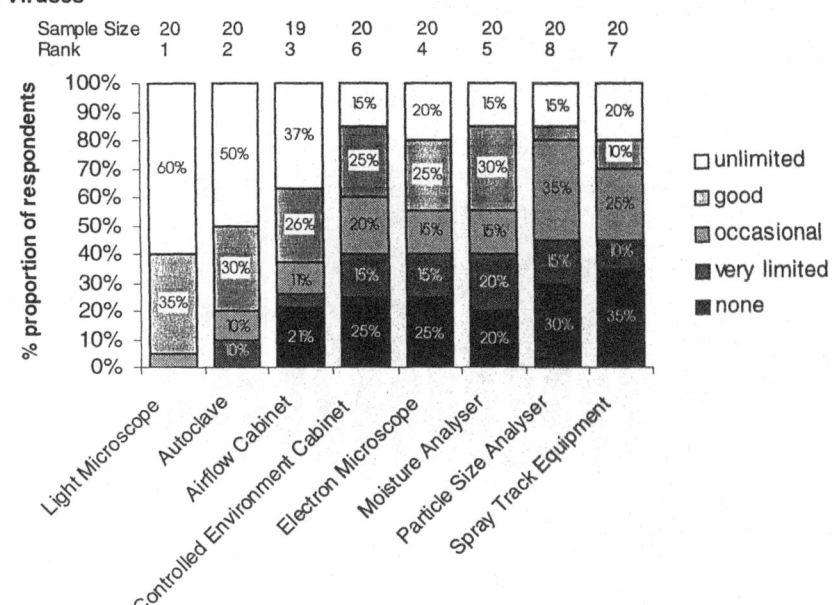

Figure 28, cont'd.

Priorities in Research Funding

a) Bacteria

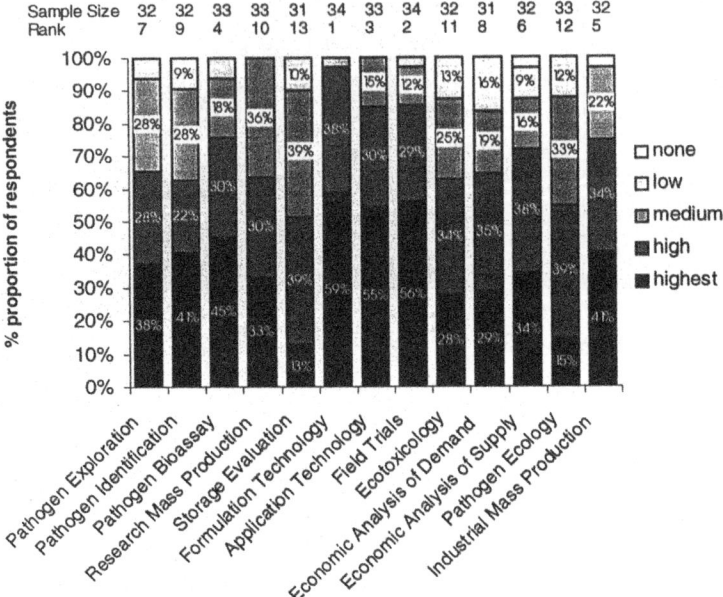

b) Fungi

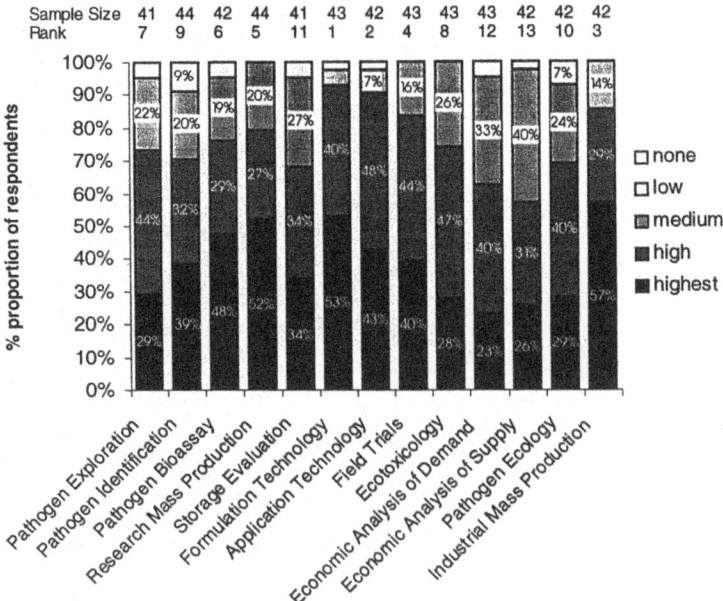

Figure 29. Respondents' priorities for funding in biopesticide research. *Cont'd.*

c) Viruses

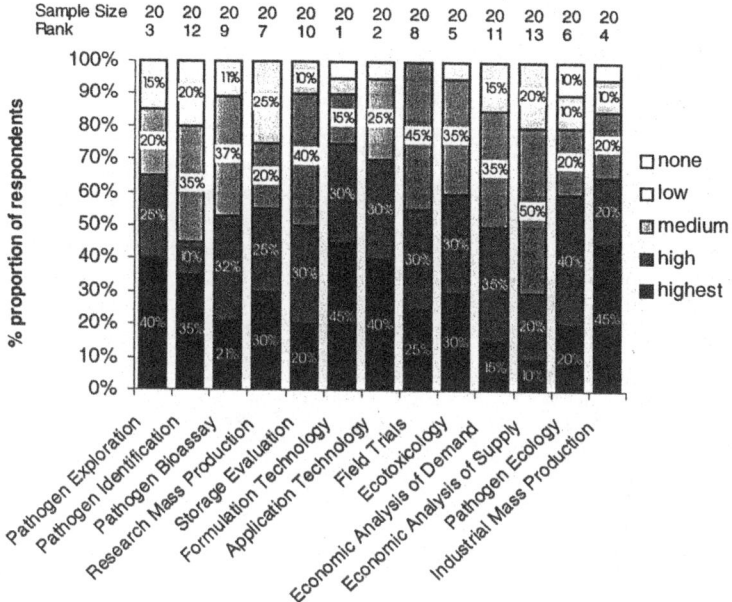

Figure 29, cont'd.

Know-how

a) Bacteria

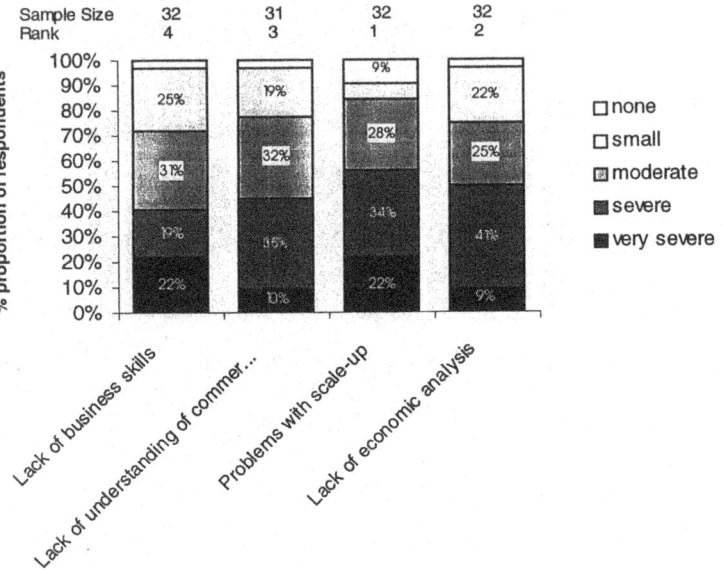

b) Fungi

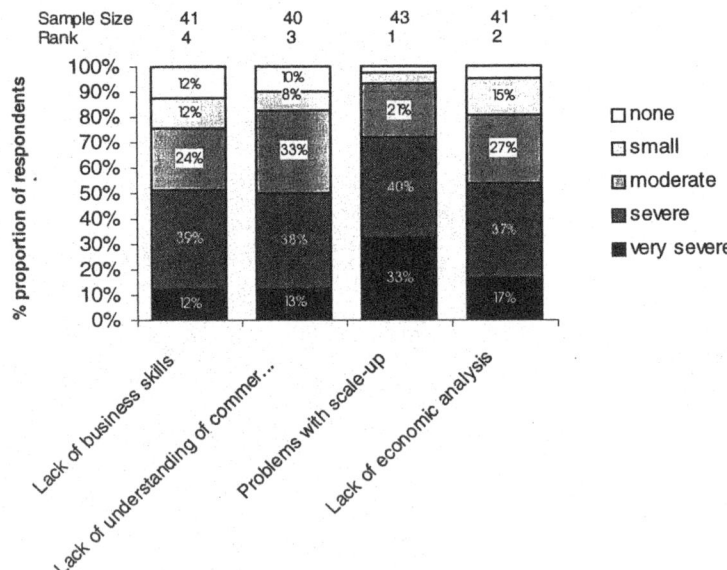

Figure 30. Respondents' evaluation of the know-how barriers to the commercialisation of biopesticides. *Cont'd.*

c) Viruses

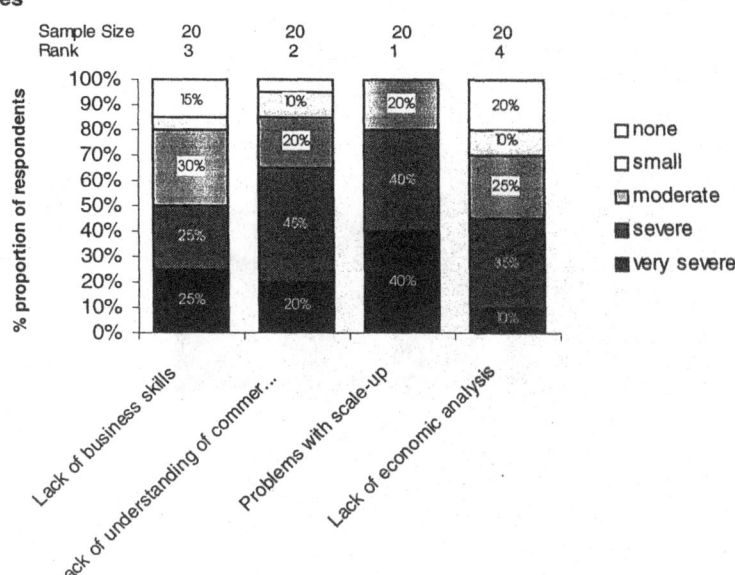

Figure 30, cont'd.

Financial and Markets

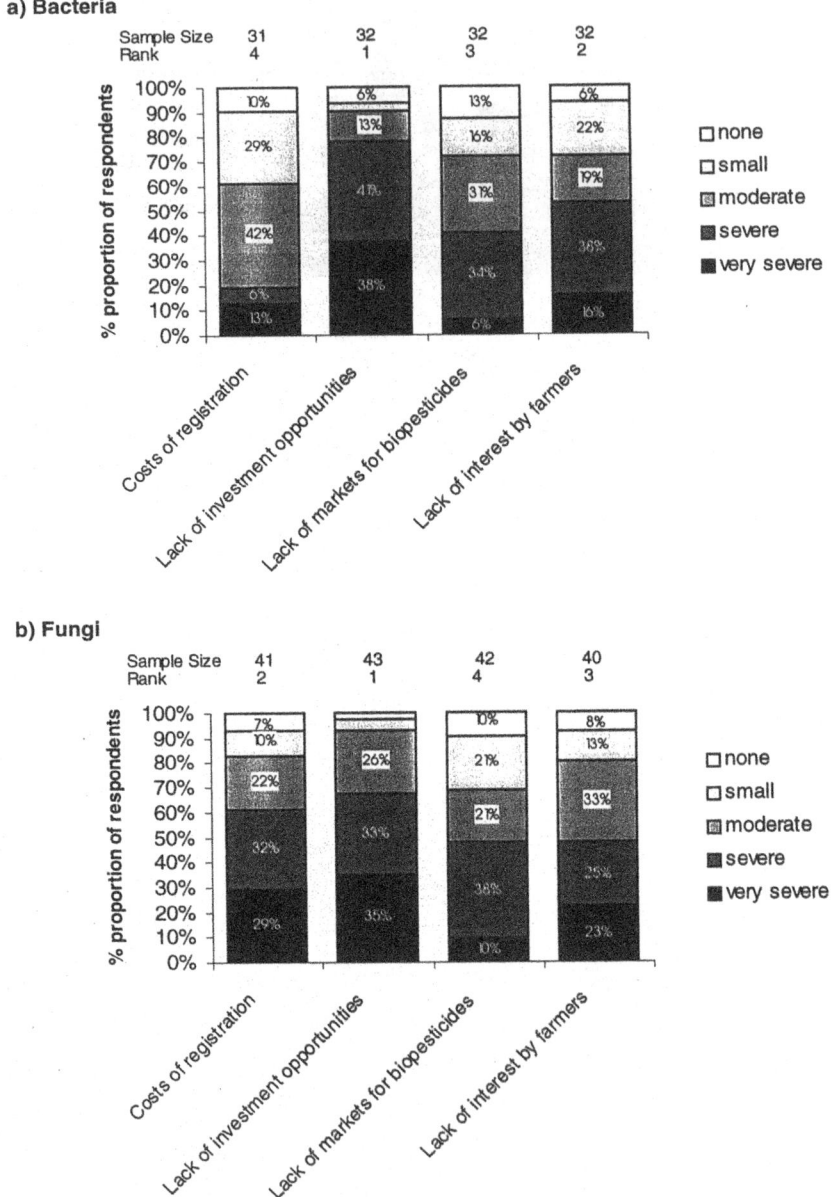

Figure 31. Respondents' evaluation of the financial barriers to the commercialisation of biopesticides. *Cont'd.*

c) Viruses

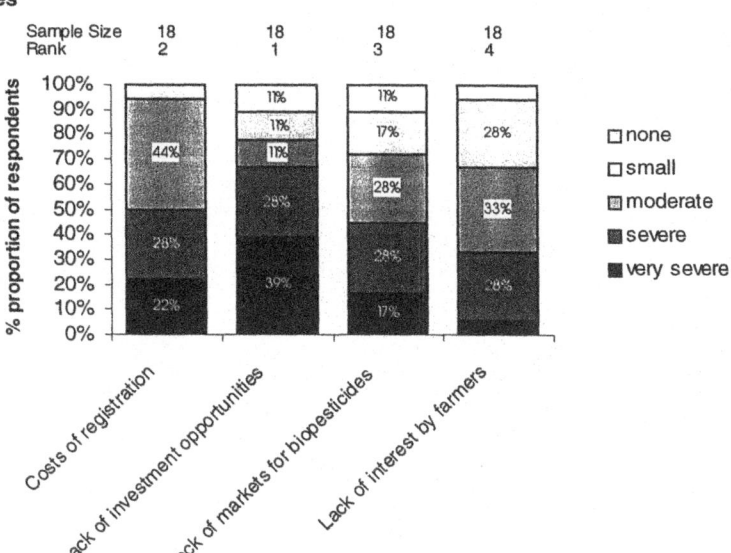

Figure 31, cont'd.

Appendix 2: Regions

Information

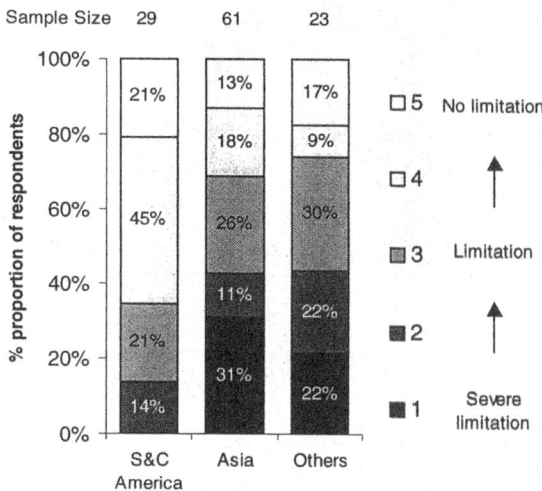

Figure 32. Availability of biopesticide information as a limitation to research.

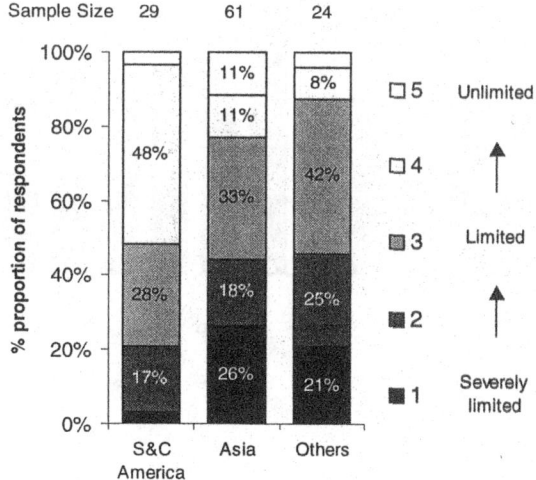

Figure 33. Respondents' own access to information on biopesticides.

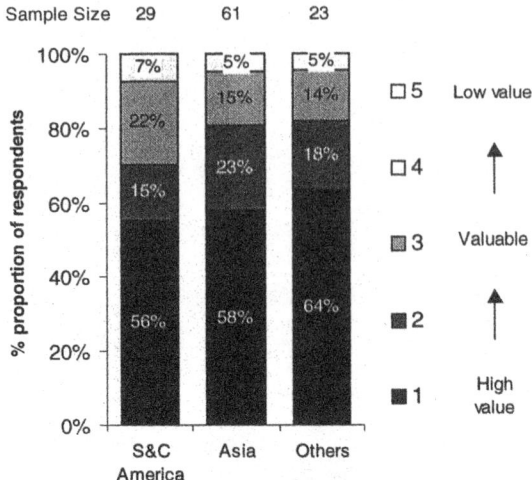

Figure 34. Value to respondents of summary information tailored to meet the needs of biopesticide R&D.

Technical expertise

a) Latin America

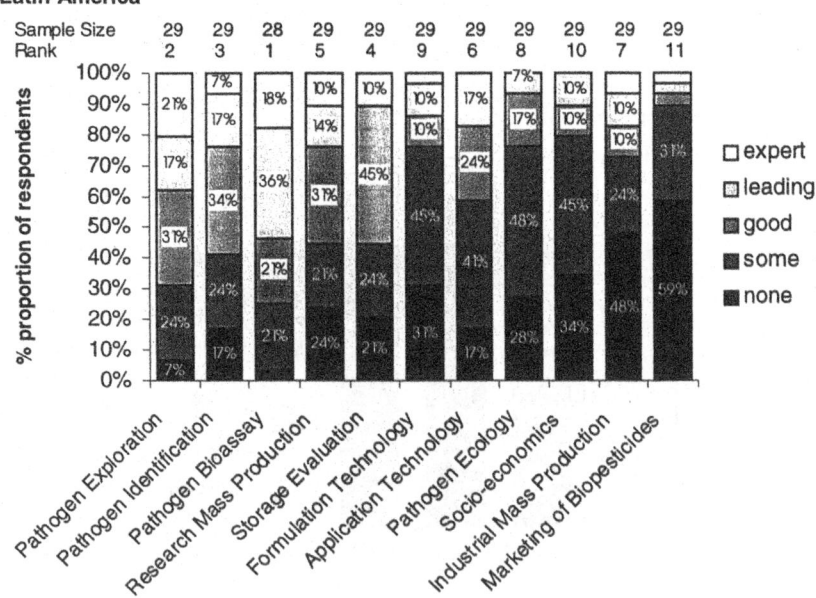

Figure 35. Respondents' ratings of their expertise in specialist areas of biopesticide development. *Cont'd.*

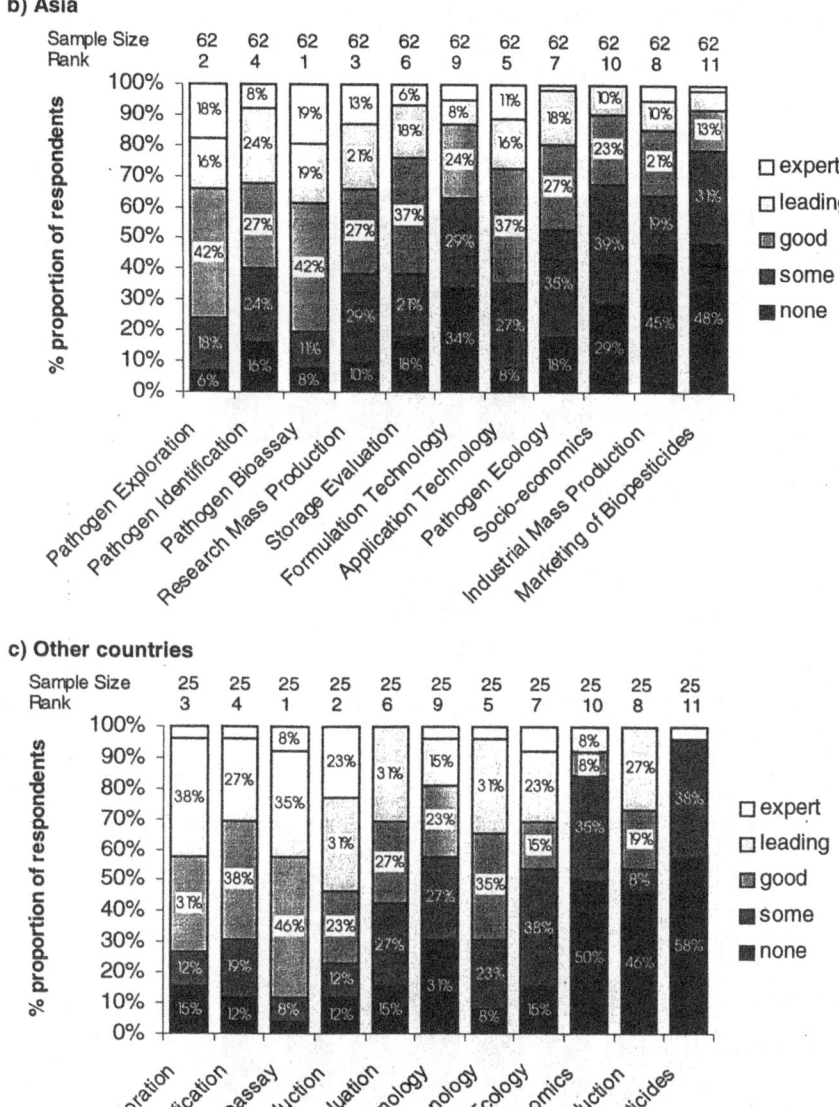

Figure 35, cont'd.

Access to Specialist Equipment

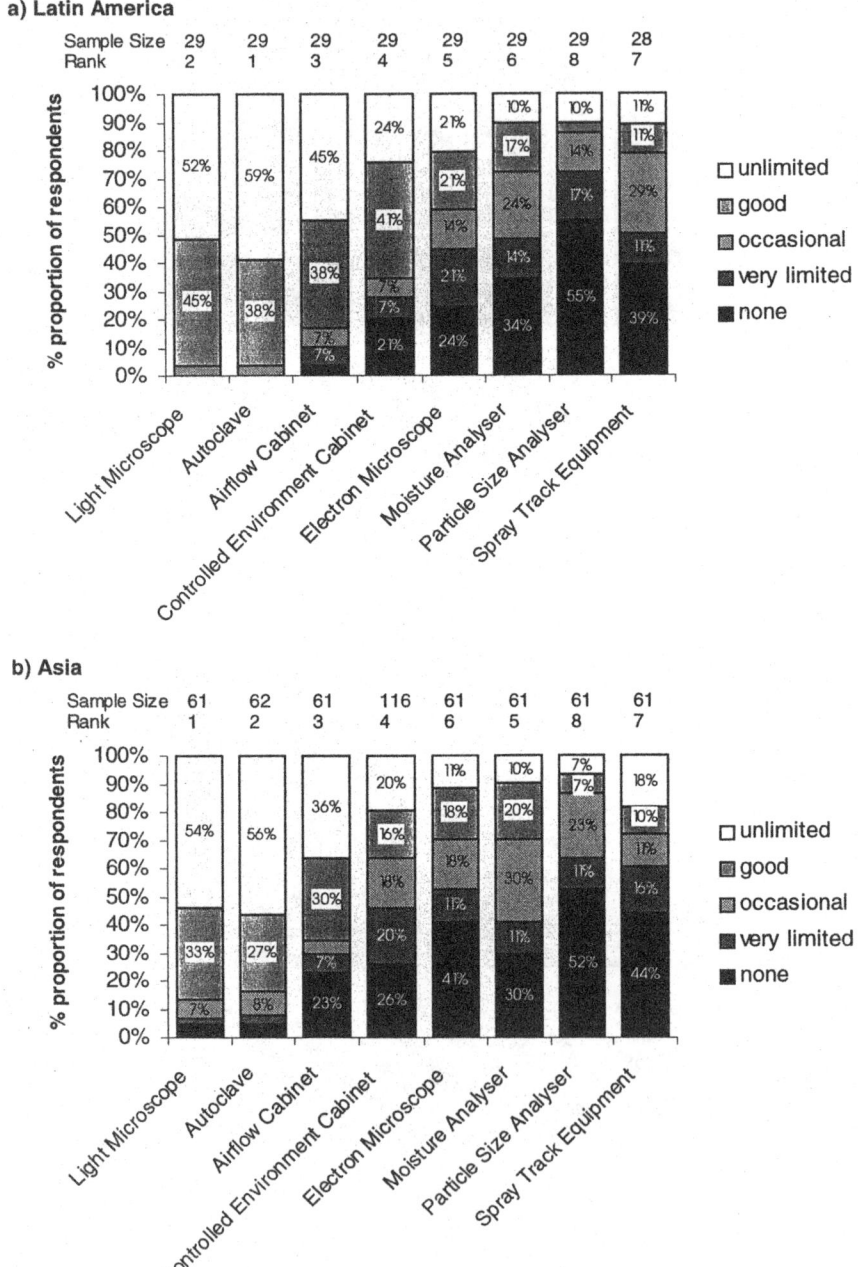

Figure 36. Respondents' access to specialist equipment. *Cont'd.*

c) Other countries

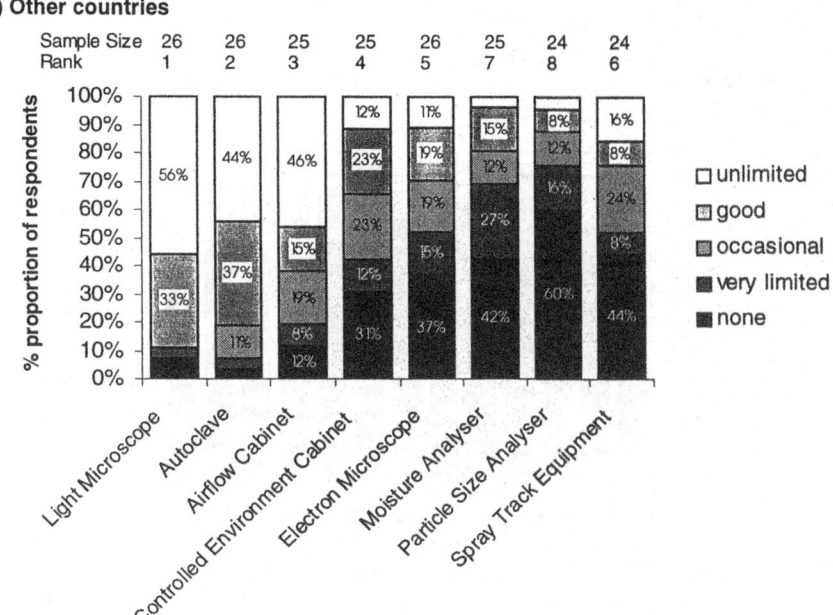

Figure 36, cont'd.

Priorities in Research Funding

a) Latin America

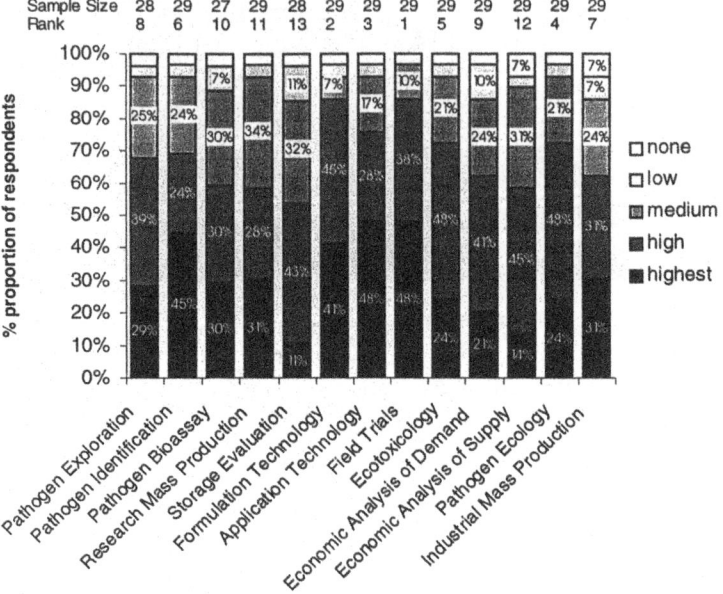

b) Asia

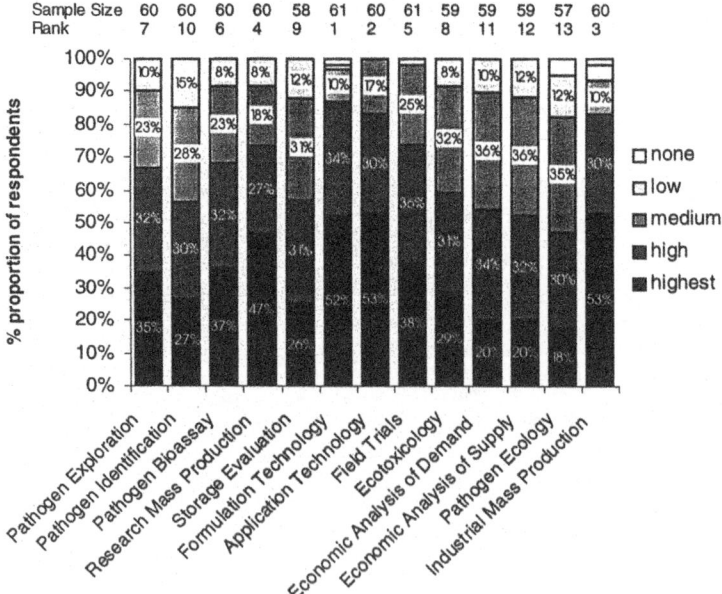

Figure 37. Respondents' priorities for funding in biopesticide research. *Cont'd.*

c) Other countries

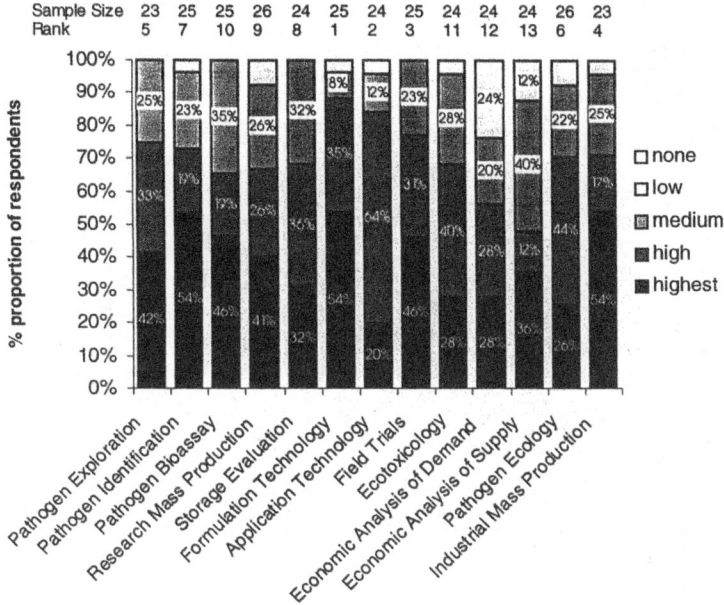

Figure 37, cont'd.

Know-how

a) Latin America

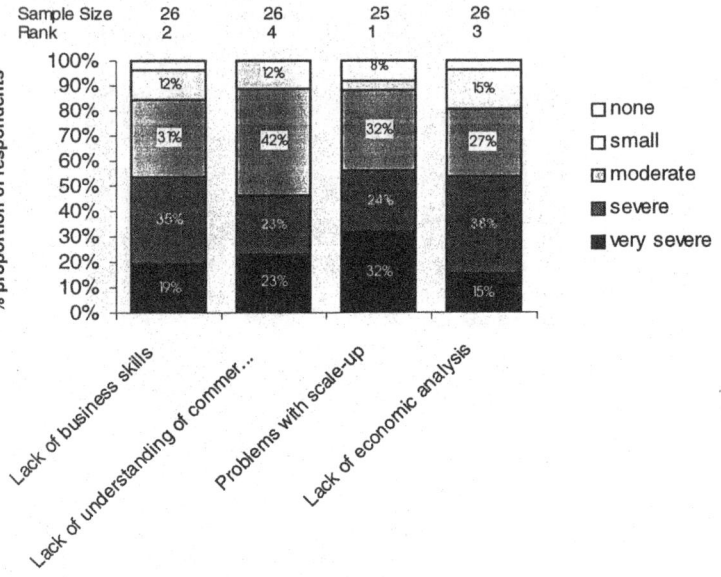

b) Asia

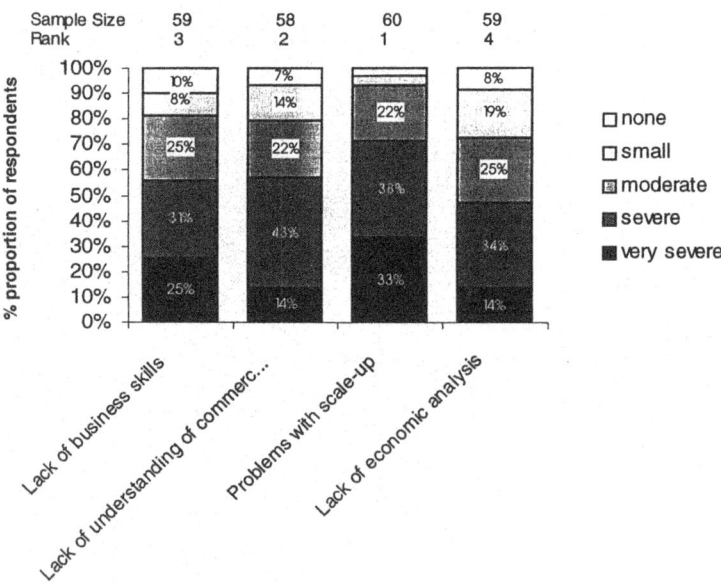

Figure 38. Respondents' evaluation of the know-how barriers to the commercialisation of biopesticides. *Cont'd.*

c) Other countries

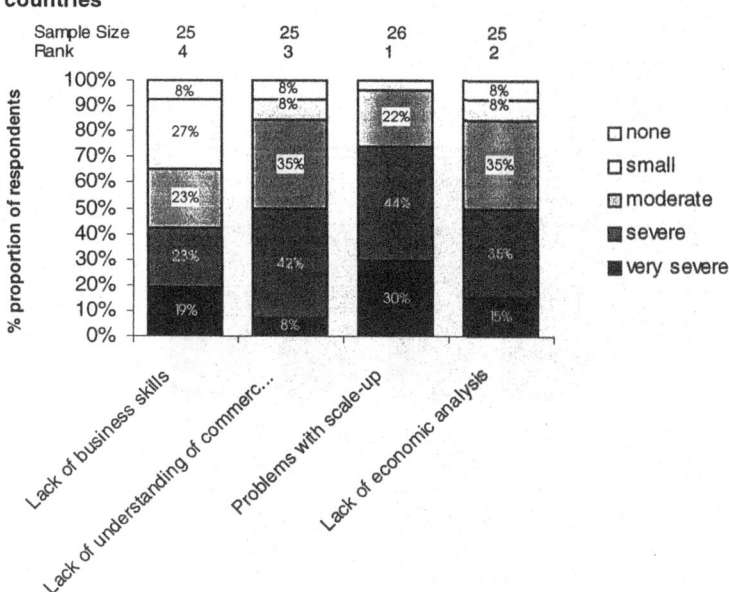

Figure 38, cont'd.

Financial and Markets

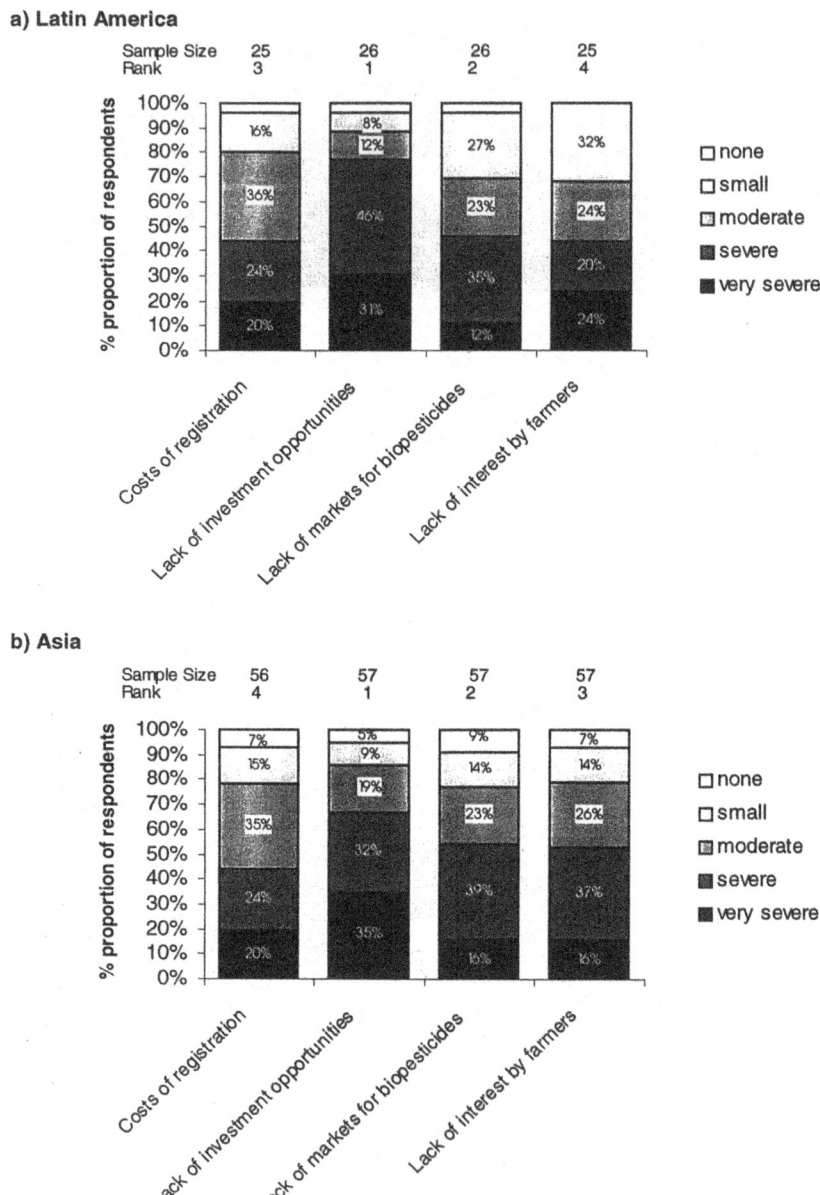

Figure 39. Respondents' evaluation of the financial barriers to the commercialisation of biopesticides. *Cont'd.*

c) Other countries

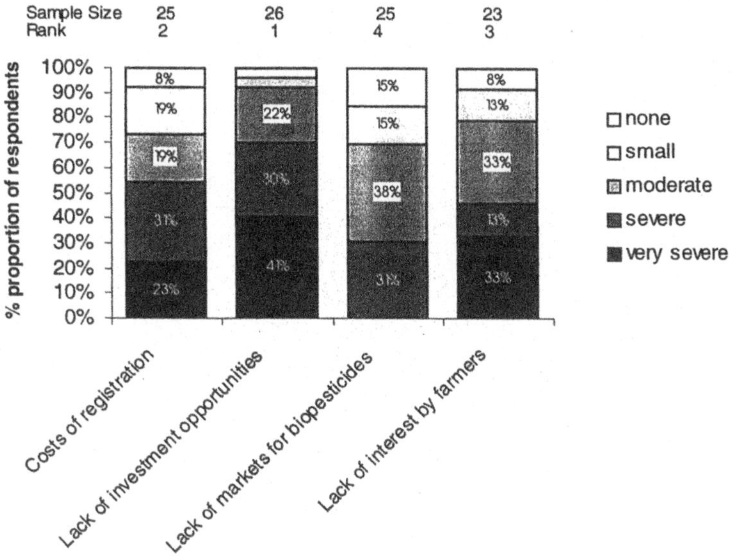

Sample Size 25 26 25 23
Rank 2 1 4 3
% proportion of respondents
100%
90%
80%
70%
60%
50%
40%
30%
20%
10%
0%
8%
19%
19%
31%
23%
22%
30%
41%
15%
15%
38%
31%
8%
13%
33%
13%
33%
Costs of registration
Lack of investment opportunities
Lack of markets for biopesticides
Lack of interest by farmers
none
small
moderate
severe
very severe

Figure 39, cont'd.

Appendix 3: The Questionnaire

Biopesticides* have enormous potential as substitutes for chemical pesticides and for use in IPM programmes but their development, commercialisation and use has not yet lived up to expectations. The purpose of this questionnaire is to help identify the constraints to successful biopesticide development around the world and the barriers that impede commercial exploitation of potential biopesticide products. The information gained from your response will be collated and used to prioritise areas for funding and assistance.

Please take a few minutes to complete all sections of the questionnaire and return by post or fax to:

Dr David Dent
Biopesticides Programme Leader
CABI Bioscience
Silwood Park
Buckhurst Road
Ascot
Berks SL5 7TA
UK

Fax: + 44 1491 829123

Thank you for your patience and time taken to complete this questionnaire. Your help is very much appreciated.

If you would like to receive a copy of the final report, please tick this box: ☐

* Biopesticides is taken to mean pesticides based on fungi, bacteria, viruses, nematodes and protozoa.

Personal Details

1. Name ....
2. Address ....

    ....

    ....

    Telephone ....

    Fax ....

    Email ....

3. Present type of employment (please tick only one):

☐ University   ☐ Research Institute

☐ Commercial Company   ☐ Regulatory Authority

☐ Other (please specify)..

General Information

4. Biopesticide Project(s) on which you currently work; please state project title(s)

....

....

....

5. Approximate total annual monetary value of your biopesticide research/development project(s) (*local currency*)

....

6. Present country of residence

....

7. Country(s) in which biopesticide research projects are undertaken

....

8. Type of pest system you work in (*if more than one, please prioritise using numbers with 1 as highest priority*):

☐ Livestock   ☐ Forestry

☐ Medical   ☐ Veterinary

☐ Urban/domestic   ☐ Agricultural crops

☐ Horticultural   ☐ Stored products

☐ Rangeland

☐ Other (please specify)..

9. Type of pest of you are confronted with in the pest system you listed as the highest priority (1) in question 8 (*if more than one type of pest then please prioritise using numbers with 1 as highest priority*):

☐ Insects ☐ Weeds

☐ Nematodes ☐ Pathogens

☐ Other (please specify) ....

10. Type of control agent you work with (*if more than one, please prioritise using numbers with 1 as highest priority*):

☐ Fungi ☐ Viruses

☐ Nematodes ☐ Bacteria

☐ Protozoa

☐ Other (please specify) ....

Constraints to Biopesticide Research and Development

11. Information (*please circle the most appropriate number in each case*)

11a. How much is a lack of access to information (international journals and books) a limitation to your project(s) in Biopesticide R&D:

Severe Limitation   *1*   *2*   *3*   *4*   *5*   *No Limitation*

11b. Would you define your own access to international and national journals and books relevant to biopesticides as:

Severely Limited   *1*   *2*   *3*   *4*   *5*   *Unlimited*

11c. Would there be value in summary information tailored to meet your needs for Biopesticide R&D:

High Value   *1*   *2*   *3*   *4*   *5*   *Low Value*

12. Technical

12a. How would you rate your expertise in each of the following specialist areas (*please circle the most appropriate number in each case*):

Level of expertise:

	None	Some	Good	Leading	Expert
	1	*2*	*3*	*4*	*5*

Pathogen exploration		1 2 3 4 5
Pathogen identification		1 2 3 4 5
Pathogen bioassay		1 2 3 4 5
Pathogen storage evaluation		1 2 3 4 5
Research mass production methods		1 2 3 4 5
Industrial mass production methods		1 2 3 4 5
Formulation technology		1 2 3 4 5
Application technology		1 2 3 4 5
Pathogen ecology		1 2 3 4 5
Socio-economics		1 2 3 4 5
Marketing (of biopesticides)		1 2 3 4 5
Other (please specify)		1 2 3 4 5

....

12b. Have you access for your own project(s) to the following specialist equipment (*please circle the most appropriate number in each case*):

Level of access available:

	None	Very Limited	Occasional	Good	Unlimited
	1	*2*	*3*	*4*	*5*

Airflow cabinet	1 2 3 4 5
Autoclave	1 2 3 4 5
Controlled environment cabinet	1 2 3 4 5
Electron microscope	1 2 3 4 5
Light microscope	1 2 3 4 5
Moisture analyser	1 2 3 4 5
Particle size analyser	1 2 3 4 5
Spray track equipment/Potter Tower	1 2 3 4 5

12c. To achieve the successful development and commercial exploitation of biopesticides, which priority should be assigned to funding research in the following areas (*please circle the most appropriate number in each case*):

Level of Priority:

Highest	*High*	*Medium*	*Low*	*None*
1	*2*	*3*	*4*	*5*

Pathogen exploration	1	2	3	4	5
Pathogen identification	1	2	3	4	5
Pathogen bioassays	1	2	3	4	5
Pathogen storage evaluation	1	2	3	4	5
Research mass production methods	1	2	3	4	5
Industrial mass production methods	1	2	3	4	5
Formulation technology	1	2	3	4	5
Application technology	1	2	3	4	5
Field trials	1	2	3	4	5
Ecotoxicology	1	2	3	4	5
Economic and socio-economic analysis of demand	1	2	3	4	5
Economic analysis of supply (production)	1	2	3	4	5
Pathogen ecology	1	2	3	4	5
Other (please specify)	1	2	3	4	5

....

Barriers to commercialisation

13. **Know-how**: Please evaluate the limitations to the potential commercialisation of biopesticides from your own project(s) in the following areas (*please circle a score between 1 (very severe) and 5 (none)*):

Limitations to commercialisation:

Very Severe	*Severe*	*Moderate*	*Small*	*None*
1	*2*	*3*	*4*	*5*

Lack of business skills	1	2	3	4	5
Lack of understanding of commercialisation	1	2	3	4	5
Problems with scale-up (from research to industrial production)	1	2	3	4	5
Lack of economic analysis	1	2	3	4	5
Other (please specify)	1	2	3	4	5

....

Other (please specify)	1	2	3	4	5

....

14. Financial and markets: Please evaluate the limitations to potential commercialisation of biopesticides from your own project(s) in the following areas (*please circle a score between 1 (very severe) and 5 (none)*):

Limitations to commercialisation:

Very Severe	*Severe*	*Moderate*	*Small*	*None*
1	*2*	*3*	*4*	*5*

Costs of registration	1	2	3	4	5
Lack of investment opportunities	1	2	3	4	5
Lack of markets for biopesticides	1	2	3	4	5
Lack of interest by farmers	1	2	3	4	5
Other (please specify)	1	2	3	4	5

....

Printed and bound by CPI Group (UK) Ltd, Croydon, CR0 4YY

20/01/2026

02038692-0002